An Introduction to Variational Calculus

Hebert Montegranario

An Introduction to Variational Calculus

Applications in Image Processing

Hebert Montegranario
Mathematics Institute
Universidad de Antioquia
Medellin, Colombia

ISBN 978-3-031-77269-6   ISBN 978-3-031-77270-2 (eBook)
https://doi.org/10.1007/978-3-031-77270-2

© The Editor(s) (if applicable) and The Author(s), under exclusive license to Springer Nature Switzerland AG 2025

This work is subject to copyright. All rights are solely and exclusively licensed by the Publisher, whether the whole or part of the material is concerned, specifically the rights of translation, reprinting, reuse of illustrations, recitation, broadcasting, reproduction on microfilms or in any other physical way, and transmission or information storage and retrieval, electronic adaptation, computer software, or by similar or dissimilar methodology now known or hereafter developed.
The use of general descriptive names, registered names, trademarks, service marks, etc. in this publication does not imply, even in the absence of a specific statement, that such names are exempt from the relevant protective laws and regulations and therefore free for general use.
The publisher, the authors and the editors are safe to assume that the advice and information in this book are believed to be true and accurate at the date of publication. Neither the publisher nor the authors or the editors give a warranty, expressed or implied, with respect to the material contained herein or for any errors or omissions that may have been made. The publisher remains neutral with regard to jurisdictional claims in published maps and institutional affiliations.

This Springer imprint is published by the registered company Springer Nature Switzerland AG
The registered company address is: Gewerbestrasse 11, 6330 Cham, Switzerland

If disposing of this product, please recycle the paper.

I would like to dedicate this book to my beloved children, Hanna and Santiago

Preface

In this book we want to make an introduction to both the classic and the most recent aspects of variational calculus. The classical aspects refer to the tradition that goes back to the time of creation of this discipline with Euler and Lagrange and that develops until our time in close relationship with physics, continuum mechanics, and other highly applicable topics. The most recent facets refer to the fact that many topics of modern technology such as computer vision, robotics, and, in particular, digital image processing can be formulated in terms of variational problems. These new problems and solutions have configured a new way of thinking about variational calculus, strongly related to algorithms and computational applications.

However, most books dealing with applications of variational calculus in computer vision or image processing assume extensive knowledge from the reader or deal with more recent frontier research results. In our approach, the reader only needs elements of Calculus, differential equations, and basic principles of algorithms and programming, but does not need previous knowledge of Variational Calculus.

This text aims to balance three elements: mathematics, applications, and computational implementation, and we want this intention to be reflected throughout the seven chapters into which it is divided. Therefore, examples and algorithms related to image processing appear in the book as early as possible.

In Chap. 1 we introduce the topic and give a motivation and general description. Chapter 2 contains the basic mathematical elements that will be applied in the rest of the book, making geometric references. Since this is intended to be a self-contained book, Chap. 3 provides the necessary elements to obtain a practical approach to the theory of distributions and Fourier analysis. As part of signal theory, image processing has evolved using the tools of Fourier methods, so it is important to include it in the text. Chapter 4 applies the above ideas to the notion of variational derivative and its application to classical problems of variational calculus. Chapter 5 shows how the most well-known computer vision problems are actually inverse problems that can be solved using variational regularization. Chapters 6 and 7 deal with some of the most well-known applications in image processing.

The first idea for the book came about a few years ago in relation with a seismic imaging research project (Colciencias-111553130555) and I realized that my personal point of view on the subject could be useful for those who need an introduction to variational methods. I would like to express my deep gratitude to Professor Cristhian Zuluaga, head of the Mathematics Institute at the University of Antioquia (Medellin, Colombia) for supporting the book project, and my colleagues Carlos Piedrahita and Jaime Valencia for the revision of its initial versions.

As for the source code used in the book, examples in Matlab are not intended to be the best version of the algorithms, as soon as we got a good image it seemed enough to us. The code applied comes mainly from the webpage of Numerical Tours: *https://www.numerical-tours.com/*. This and other resources can be found in the website of the book.

Medellin, Colombia
August 2024

Hebert Montegranario

Contents

1 Introduction 1
   1.1   Variational Calculus 1
   1.2   The Classical Problem of Variational Calculus 3
   1.3   Image Representation 5
   1.4   Variational Methods in Image Processing 10

2 Mathematical Preliminaries 13
   2.1   The Space $\mathbb{R}^N$ 13
   2.2   Matrices $\mathbb{R}^{m \times n}$ 16
   2.3   Derivatives 19
   2.4   Taylor's Theorem 23
   2.5   Optimization in $\mathbb{R}^N$ 25
   2.6   Some Integral Properties 27
   2.7   Linear Spaces 29
   2.8   Function Spaces 29
   2.9   Analysis on Normed Spaces 31
   2.10 Inner Product and Hilbert Spaces 36

3 Linear Operators and Functionals 41
   3.1   Linear Operators 41
   3.2   Linear Functionals 43
   3.3   Continuous Linear Functionals 45
   3.4   Distributions 46
   3.5   Test Functions 49
   3.6   Derivatives of Distributions 53
   3.7   Fourier Analysis 57
   3.8   Convolution and Fourier Transform in $\mathbb{R}$ 60
   3.9   Convolution and Fourier Transforms in $\mathbb{R}^N$. 65

4 Minimization of Functionals 75
   4.1   The Gâteaux Variation 75

4.2	Local Extrema For Differentiable Functionals	78
4.3	Geometry of Surfaces	83

5 Inverse Problems and Variational Regularization ... 89
5.1	Linear Inverse Problems	89
5.2	Image Deblurring as an Inverse Problem	90
5.3	Regularization Methods	92
5.4	Functionals with Physical Interpretation	94
5.5	Total Variation	98
5.6	Convexity and Convex Functionals	102

6 Variational Curve and Surface Reconstruction ... 107
6.1	Polynomial Interpolation	107
6.2	The Limitations of Polynomial Approximation	109
6.3	Spline's Variational Theory	110
6.4	Regularization and Smoothing Splines	115
6.5	Extensions to Higher Dimensions	118
6.6	Thin Plate Spline and Scattered Data Interpolation	120

7 Variational Image Feature Extraction ... 127
7.1	The Mumford-Shah Functional	127
7.2	Variational Level Set Methods	132
7.3	Active Contours	134
7.4	The Chan-Vese Model for Image Segmentation	138

References ... 141

Index ... 147

Chapter 1
Introduction

This introductory chapter provides a historical and conceptual foundation for variational methods in image processing. It begins with an exploration of classical variational calculus, tracing its origins to the ideas of Maupertuis, Euler, and Lagrange, and their connection to Enlightenment philosophy's quest for optimality and minimal principles in nature [67, 75, 124].

The discussion transitions to a mathematical introduction to the gradient of a functional, a key concept in variational methods, and the classical problem of minimizing functionals of the form $J[u] = \int F(x, y, y')dx$ which serves as the cornerstone for many optimization problems in calculus of variations.

Next, the chapter contrasts discrete and continuous representations of images, providing a framework for understanding how variational methods are applied in both settings. The chapter also introduces the fundamental tasks of imaging, such as denoising, deblurring, and other restoration techniques.

Finally, an introduction to variational methods in image processing is presented, highlighting key models like the ROF model for denoising and the Mumford-Shah model for image segmentation. This sets the stage for the deeper exploration of these methods throughout the book.

1.1 Variational Calculus

The evolution of variational calculus follows a long tradition that comes from the times of Maupertuis [4, 9] and the enlightenment philosophers about the harmony of universe until the most recent applications of variational calculus in different problems of science and technology.

This evolution has been very enriching and complex, it has followed many paths and greatly influenced all of mathematics, specially functional analysis. There are many works that describe this process [4, 51, 99].

In this work we are particularly interested in following the common thread that begins with the Euler-Lagrange equation up to state of the art applications in computer vision. We want to show variational calculus as a sub-chapter of the Differential Calculus in Banach spaces applied to the optimization of functionals. In this way, the well known results for the optimization of functions in one or several variables can be extended naturally to the problems of variational calculus. In particular, we are interested in showing how the notion of **gradient of a functional** is defined by Gateaux derivative. As a consequence, algorithms can be translated from $\mathbb{R}^N$ to general Banach spaces, in particular the gradient descent algorithm.

Given the function $f : \mathbb{R}^N \to \mathbb{R}$ continuously differentiable and bounded from below the gradient descent algorithm finds a numerical approximation to the problem

$$\min_{\mathbf{x} \in \mathbb{R}^N} f(\mathbf{x})$$

by an iteration of the form

$$\mathbf{x}_{k+1} = \mathbf{x}_k - t_k \nabla f(\mathbf{x}_k)$$

where ∇f is the gradient of the function f. In variational calculus, the unknown quantity is no longer a vector $\mathbf{x}$, but a function u, and what is minimized is no longer a function $f(\mathbf{x})$, but a functional $\mathscr{J}[u]$.

Having a well defined derivative $\delta \mathscr{J}(u; \varphi)$ of a functional $\mathscr{J}$ we can formulate an analog of the gradient method. Of particular importance is the case where an inner product can be introduced into the domain of the functional $\mathscr{J}$, such as $\langle u, v \rangle = \int uv$. In this case, the Gâteaux derivative $\delta \mathscr{J}(u; \varphi)$ of the functional $\mathscr{J}$ at the point u and the gradient $\nabla \mathscr{J}$ are defined such a manner that satisfy

$$\lim_{\varepsilon \to 0} \frac{\mathscr{J}(u + \varepsilon \varphi) - \mathscr{J}(\varphi)}{\varepsilon} = \langle \nabla \mathscr{J}(u), \varphi \rangle = \int (\nabla \mathscr{J}) \varphi \qquad (1.1)$$

So the Gradient method is again

$$u_{k+1} = u_k - \alpha_k \nabla \mathscr{J}[u_k] \qquad (1.2)$$

The notation $\nabla \mathscr{J}[u]$ is really adequate for the variational derivative. Its properties are similar to the finite dimensional case. In both cases, the equation $\nabla \mathscr{J}[u] = 0$ is a necessary condition, but is not sufficient for a local minimum. Sufficient conditions can be obtained from the so called *second variation* of $\mathscr{J}[u]$, which is equivalent to the Hessian of a function. Such as we will see later, the notion of **convexity** also plays an important role in the statement of necessary and sufficient conditions for optimality.

Depending on the context, literature, and the area we are working on, the Gâteaux derivative may have different names and interpretations. Sometimes the left hand side of Euler-Lagrange equation (1.8) is called **functional derivative**. Some authors [9, 37, 127] prefer to say **variational derivative** and write

$$\delta \mathscr{J}_h(u) = \int \frac{\delta J}{\delta u} \varphi(x) dx$$

meaning that $\frac{\delta J}{\delta u}$ is the functional derivative of the functional $\mathscr{J}$.

According to Vainberg [118] this concept of the gradient (1.1) of a functional is first encountered in Courant-Hilbert [27] (pp. 222–224) without a precise definition. The first formal definition of the gradient of a functional is due to Golomb [52] who considered functionals differentiable in the sense of Frechet [113].

1.2 The Classical Problem of Variational Calculus

In this introduction we briefly describe the classical approach to variational calculus in different problems of science and technology. The first methods were designed by Euler and Lagrange by the equation which bears their names. These ideas were the ferment of great discoveries and long discussions between the great philosophers of the eighteenth century. They are commonly expressed in terms of optimization problems on integrals [15, 46, 48, 59].

Several problems of mathematics deal with mappings or operators acting on functions. Those operators $\mathscr{J}[\cdot]$ that assigns real or complex numbers to functions are called **functionals**; the term **operator** is usually applied to transformations that carry functions into functions. In particular, a wide family of problems are given by functionals in the form

$$\mathscr{J}[y] = \int_a^b F(x, y, y') dx \qquad (1.3)$$

Example Although it is evident that the shortest path between two points in the plane is a line segment, this property can be formulated as the variational problem of finding the path $y(x)$ between two points $A(x_0, y_0)$, $B(x_1, y_1)$ in the plane for which the arc length integral

$$\mathscr{J}[y] = \int_{x_0}^{x_1} \sqrt{1 + y'^2}\, dx \qquad (1.4)$$

attains its minimum value. □

Fig. 1.1 The approach of Lagrange for obtaining a necessary condition for an extremum of the functional

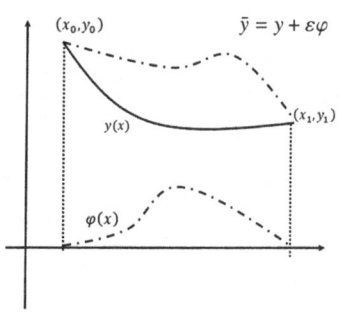

The Euler-Lagrange Equation

In order to find the minimum of the functional (1.3), assume the optimum path is $y(x)$. The method consists in compare $\mathscr{J}$ for the optimum path with that obtained from neighboring paths. Assuming everything is sufficiently differentiable for our purposes, consider the set of functions φ such that $\varphi(x_0) = \varphi(x_1) = 0$ (Fig. 1.1) and if ε is a small parameter,

$$\bar{y} = y + \varepsilon\varphi \tag{1.5}$$

with $\bar{y}' = y + \varepsilon\varphi'$, represents a one parameter family of admissible functions that satisfy the boundary conditions, such that $\bar{y}(x_i) = y(x_i)$, $i = 1, 2$.

Observe that we can define an univariate function $\Phi : \mathbb{R} \to \mathbb{R}$, such that $\varepsilon \in \mathbb{R}$ and

$$\Phi(\varepsilon) := \mathscr{J}[y + \varepsilon\varphi]$$

$$\Phi(\varepsilon) = \int_{x_0}^{x_1} F(x, y(x) + \varepsilon\varphi(x), y'(x) + \varepsilon\varphi'(x))dx \tag{1.6}$$

This expression reduces the problem to a question in one-variable calculus; so the minimum is found by making $\Phi'(0) = 0$ then

$$\Phi'(0) = \lim_{\varepsilon \to 0} \frac{\Phi(\varepsilon) - \Phi(0)}{\varepsilon} = \lim_{\varepsilon \to 0} \frac{\mathscr{J}[y + \varepsilon\varphi] - J[y]}{\varepsilon}$$

$$= \frac{\partial}{\partial \varepsilon} \mathscr{J}[y + \varepsilon\varphi]\bigg|_{\varepsilon=0}$$

we call this limit, the variation of y in the direction of φ and now is clear that the minimum of the integral is found with

$$\frac{\partial \Phi}{\partial \varepsilon}\bigg|_{\varepsilon=0} = 0$$

1.3 Image Representation

We shall assume below that the function $F(x, y, y')$ possesses all the necessary derivatives. Then applying the chain rule

$$\frac{\partial}{\partial \varepsilon} F(x, \bar{y}, \bar{y}') = \frac{\partial F}{\partial x} \overbrace{\frac{\partial x}{\partial \varepsilon}}^{=0} + \frac{\partial F}{\partial \bar{y}} \overbrace{\frac{\partial \bar{y}}{\partial \varepsilon}}^{=\varphi} + \frac{\partial F}{\partial \bar{y}'} \overbrace{\frac{\partial \bar{y}'}{\partial \varepsilon}}^{=\varphi'}$$

$$\frac{\partial \Phi}{\partial \varepsilon} = \int_{x_0}^{x_1} \frac{\partial}{\partial \varepsilon} F(x, \bar{y}, \bar{y}') dx = \int_{x_0}^{x_1} \frac{\partial F}{\partial \bar{y}} \varphi + \frac{\partial F}{\partial \bar{y}'} \varphi'$$

Now we apply integration by parts and $\varphi(x_0) = \varphi(x_1) = 0$

$$\int_{x_0}^{x_1} \frac{\partial F}{\partial y'} \varphi' dx = \overbrace{\left[\frac{\partial F}{\partial y'} \varphi(x) \right]_{x_0}^{x_1}}^{=0} - \int_{x_0}^{x_1} \frac{d}{dx} \left(\frac{\partial F}{\partial y'} \right) \varphi \, dx$$

$$= - \int_{x_0}^{x_1} \frac{d}{dx} \left(\frac{\partial F}{\partial y'} \right) \varphi \, dx$$

Then

$$\left. \frac{\partial \Phi}{\partial \varepsilon} \right|_{\varepsilon=0} = \int_{x_0}^{x_1} \left[\frac{\partial F}{\partial y} - \frac{d}{dx} \left(\frac{\partial F}{\partial y'} \right) \right] \varphi = 0 \qquad (1.7)$$

Finally, as we will see later, $\int f \varphi = 0 \quad \forall \varphi$ implies $f = 0$ then we have as necessary condition for a minimum, the famous Euler-Lagrange equation [28, 29, 36]

$$\frac{\partial F}{\partial y} - \frac{d}{dx} \left(\frac{\partial F}{\partial y'} \right) = 0 \qquad (1.8)$$

1.3 Image Representation

Image processing [58, 88] is a very important area of computer vision and artificial intelligence. Of all the external information our brain receives, visual information makes up the largest part; in fact, about 80% of all that information. In this way, human beings can detect and recognize a large number of objects regardless of their appearance and conditions [77, 97]. This tremendous capacity, of which we are not aware, is due to the fact that we have a powerful system of visual perception and a great memory. This gives us an ability to learn and remember what we see and feel. We can recognize scenes and adapt to new environments. Computer vision seeks mechanisms for computers to be able to perceive, interpret and understand what surrounds us in the same way as humans do.

In imaging, there are essentially two different approaches: the discrete setting and the continuous setting. Each approach offers unique perspectives and tools for handling image data, with distinct implications for analysis, processing, and interpretation. In the discrete setting, images are represented as a grid of pixels. Each pixel is treated as a distinct entity, holding a specific intensity value or color. Images are often stored as matrices or arrays, with each element corresponding to a pixel value. This representation is common in digital imaging where images are captured, stored, and processed in a pixel-by-pixel manner.

Discrete methods are widely used in applications such as image compression, digital photography, and medical imaging, where the pixel-based structure is inherent to the data acquisition process. This approach provides compatibility with digital devices and directly matches the way images are captured by digital cameras and sensors. Algorithms designed for discrete data are often simpler to implement and computationally efficient.

In the *continuous approach*, images are modeled as continuous functions defined over a continuous spatial domain. Each point in this domain has an associated intensity value. This approach uses mathematical functions and practically all the tools associated with that concept. Continuous methods are used in theoretical analysis, computer vision, and in scenarios where the underlying processes are naturally continuous. Allows the application of advanced mathematical tools and theories, such as calculus of variations and partial differential equations (PDEs), leading to more sophisticated models and algorithms. Solving continuous models often requires complex numerical methods and can be computationally intensive [26]. Continuous models ultimately need to be discretized for implementation on digital computers, which can introduce approximation errors.

Historically, the discrete approach dominated the early days of digital imaging due to the limitations of computing power and the nature of digital image acquisition. With advances in computational methods and theory, the continuous approach gained prominence, particularly in academic and theoretical research. Today, the integration of both approaches reflects a mature understanding of their complementary roles in advancing image processing and computer vision.

Digital Image Representation

In the problems of image processing and analysis an image is a continuous object defined on a rectangular domain $\Omega =]x_0, x_1[\times]y_0, y_1[\subset \mathbb{R}^2$ and mathematically represented as an element u in a function space $\mathcal{U}$ such that $u : \Omega \to \mathbb{R}$. Here the function value $u(x, y)$ represents the intensity of the image in the pixel $\mathbf{x} = (x, y)$ of the image domain Ω.

The result of sampling and quantization of the image u is a $M \times N$ matrix $U_{jk} = u(j, k)$ of real numbers.

1.3 Image Representation

Fig. 1.2 The surface $z = \sin x$ (left), seen as an image $u = |\sin x|$ in gray scale (right)

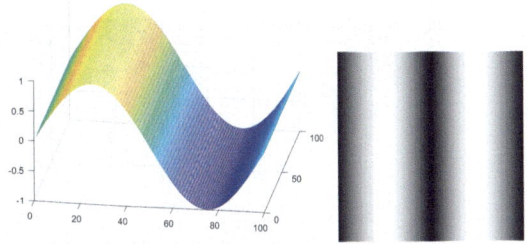

$$u(x, y) = \begin{bmatrix} u(1,1) & u(1,2) & \cdots & u(1,N) \\ u(2,1) & u(2,2) & \cdots & u(2,N) \\ \vdots & \vdots & \ddots & \vdots \\ u(M,1) & u(M,2) & \cdots & u(M,N) \end{bmatrix}$$

The brightness values of pixels in a digital image depend on the color model used to represent the image. The most common color models are RGB (Red, Green, Blue) and grayscale [126].

In a **grayscale image** (Fig. 1.2), each pixel is represented by a single intensity value ranging from 0 to 255 in an 8-bit image. Here: 0 corresponds to black (no intensity), 255 corresponds to white (maximum intensity), and Values in between represent varying shades of gray.

In an **RGB image**, each pixel is represented by three intensity values (one for each color channel: Red, Green, Blue). Each channel typically has a range from 0 to 255 in an 8-bit image. A pixel's brightness or luminance is often computed as a weighted sum of its RGB values.

In **images with higher bit depth** (e.g., 16-bit or 32-bit), the intensity values can have a wider range, providing more precision for representing color or grayscale information.

It's important to note that these conventions and ranges are typical but not universal. Some images might use different color spaces, bit depths, or normalization methods. Additionally, the perceived brightness of an image can be influenced by factors such as gamma correction, color profiles, and display characteristics.

When working with digital images it is crucial to understand the color space and bit-depth of the images you are dealing with, as this knowledge is fundamental for accurate image processing and analysis.

The Basic Tasks of Imaging

We are interested in images not only for the visual or artistic pleasure they provide us but also for the information we can extract from them (Fig. 1.3). This information may refer to attributes of the image that we cannot perceive or that are not

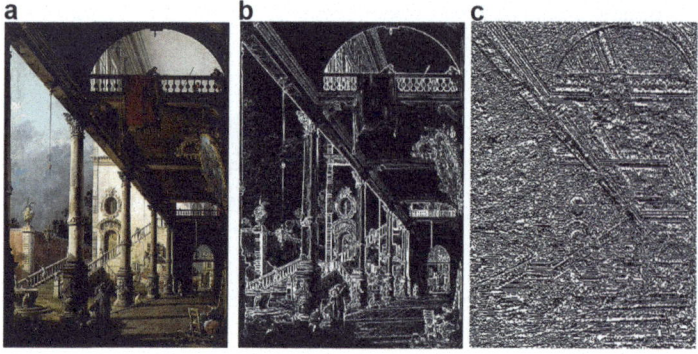

Fig. 1.3 We are interested in images not only for the visual or artistic pleasure they provide us but also for the information we can extract from them. (**a**) Canaletto's *Prospettiva con portico* (**b**) Segmentation (**c**) Segmentation by gradient

our expertise. During the last decades their applications have increased and today are important part of big data and artificial intelligence applications.

There is a large number of processes that can be carried out on an image, depending on each particular application, however there are some basic tasks that have been established as the most common.

Denoising and deblurring are two different image processing techniques that aim to improve the quality of an image. The main difference between the two is the type of noise or degradation that they address. Denoising is a technique used to remove noise from an image. **Noise** is any unwanted random variation or interference that affects the quality of an image. Noise can be caused by various factors such as poor lighting conditions, sensor limitations, or transmission errors. **Denoising** aims to remove this noise while preserving the important image details.

Deblurring, on the other hand, is a technique used to remove blur from an image. **Blur** is caused by the motion of the camera or the object being photographed, which causes the image to appear fuzzy or out of focus. Deblurring aims to restore the sharpness and clarity of the image. In summary, denoising is used to remove noise, while deblurring is used to remove blur. Both techniques can be used in combination to improve the quality of an image that suffers from both noise and blur.

Segmentation refers to the process of dividing an image into meaningful and homogeneous regions or segments based on certain characteristics, such as color, intensity, texture, or other visual properties. The goal of segmentation is to simplify the representation of an image and make it more meaningful and easier to analyze.

The primary purposes of image segmentation include:

Object Recognition Segmentation helps identify and delineate individual objects or regions within an image. This is crucial for computer vision tasks such as object detection, tracking, and recognition.

1.3 Image Representation

Image Understanding Segmenting an image into meaningful regions contributes to a better understanding of its content. It allows for the extraction of relevant information from different parts of the image.

Feature Extraction Segmentation aids in the extraction of features from specific regions of an image. These features can be used for subsequent analysis or classification tasks.

Segmentation involves assigning each pixel of an image to a specific category. The segmentation process makes it possible to simplify the representation of a digital image, partitioning it into regions with defined characteristics that allow locating certain objects or borders within the image.

Edge detection in image processing refers to the process of identifying and locating sharp discontinuities or boundaries in an image. These boundaries often correspond to significant changes in intensity, color, or texture and are crucial for understanding the structure of the objects in the image.

The primary goal of edge detection is to highlight or mark the areas in an image where the intensity changes sharply, representing transitions between different objects or regions. Edges are essential features in image analysis and computer vision tasks because they often correspond to object boundaries, which are crucial for tasks such as object recognition, image segmentation, and scene understanding.

Inpainting refers to the process of reconstructing lost or damaged parts of an image, typically by filling in the missing regions with visually plausible information. The goal is to make the inpainted areas seamlessly blend with the surrounding content, giving the impression that the image never had missing or damaged portions.

There are several scenarios where inpainting can be applied:

Restoration of Damaged Images Inpainting can be used to restore images that have been corrupted or damaged, such as old photographs with scratches, stains, or other imperfections.

Object Removal Inpainting can be used to remove unwanted objects or elements from an image. The algorithm fills in the regions left vacant after removing an object in a way that is visually coherent with the surrounding content.

Image Editing and Manipulation Inpainting tools are commonly used in image editing software to fill in areas after removing unwanted objects, text, or other elements. It allows users to modify images without leaving obvious traces of the changes.

Hole Filling in Computer Vision In computer vision applications, inpainting may be used to fill in gaps or holes in images that occur due to occlusions or missing data.

Image **registration** refers to the process of aligning two or more images of the same scene, taken at different times, from different viewpoints, or by different sensors. The goal is to find a transformation that maps the pixels of one image to the corresponding pixels in another, making them spatially consistent. Image registration is a fundamental step in various applications, enabling the comparison, analysis, and integration of information from multiple images. Vision research is generally classified into different levels: low, middle, and high.

Low-level vision refers to the initial processing of raw visual input, primarily focusing on basic visual features such as edges, corners, colors, textures, and motion. Examples of tasks associated with low-level vision include edge detection, motion detection, color segmentation, and texture analysis. These tasks are typically characterized by their reliance on local image properties.

Middle-level vision involves the integration and interpretation of low-level visual features to extract more meaningful structures and relationships within the visual scene. For example contour grouping, figure-ground segregation, object recognition, depth estimation, and scene parsing. These tasks often require more sophisticated algorithms and may involve contextual information and global scene analysis. Unlike low-level vision tasks, which operate on individual image elements, middle-level vision tasks involve analyzing relationships and interactions between different visual elements to derive higher-order perceptual representations.

High-level vision encompasses the highest level of visual processing, involving cognitive and semantic understanding of the visual scene. This level of processing goes beyond the analysis of individual objects and focuses on interpreting the meaning, context, and intent behind the visual information.

Tasks associated with high-level vision include object categorization, scene understanding, visual reasoning, semantic segmentation, and visual attention. These tasks require complex cognitive processes such as memory, attention, inference, and decision-making. High-level vision is closely related to other cognitive processes such as language, memory, and reasoning, and it plays a crucial role in higher-order cognitive functions such as perception, comprehension, and action planning.

While low, middle, and high-level vision are often conceptualized as distinct stages, it's important to recognize that they are interconnected and interdependent. Information flows bidirectionally between these levels, with higher-level processing influencing lower-level processing and vice versa.

1.4 Variational Methods in Image Processing

Variational methods have become in recent years the most effective for modeling and solving computer vision problems and in particular image processing [119]. It is important to note that in these applications the functional to be optimized usually appears in a regularization format, that is, it is required to find a stable approximation to a solution of an operator equation

$$\mathcal{K}(u) = f$$

1.4 Variational Methods in Image Processing

where we have available noisy data f^δ of the exact solution f. Regularization methods propose to solve the problem [7, 38, 92, 115] by minimizing the functional

$$\mathcal{T}_{\lambda, f^\delta}(u) := \rho(\mathcal{K}(u), f^\delta) + \lambda \mathcal{R}(u) \tag{1.9}$$

For example, in the **denoising** problem the image f^δ is assumed to have additive noise (zero mean Gaussian noise η) in the model

$$f^\delta = u + \eta$$

where the task of recovering $u(x, y)$ from f^δ is an inverse problem that can be solved by minimizing a functional (1.9) that may have the form

$$\min_u \mathcal{T}(u) = \int_\Omega (u - f^\delta)^2 dy\, dx + \lambda \mathcal{R}(u)$$

where $\lambda > 0$ and $\mathcal{R}$ is called the Regularization functional.

Two of the most representative regularization models in image processing have been:

1. The ROF (Rudin-Osher-Fatemi) model for image denoising and
2. The MS (Mumford-Shah) model for image segmentation

In the ROF problem [102] the regularization functional is the total variation

$$\mathcal{R}(u) = TV(u) = \int_\Omega \|\nabla u\| dy\, dx$$

where, Ω is the image domain, usually taken as a bounded domain in $\mathbb{R}^N$ with Lipschitz boundary. In applications Ω is simply a rectangle in $\mathbb{R}^2$ modeling the computer screen. The function $u : \Omega \to \mathbb{R}$ represents the given observed image.

In the MS problem, $\Omega \subset \mathbb{R}^2$ is a bounded open connected set, γ a compact curve in Ω and $f : \Omega \to \mathbb{R}$ is a given image. Mumford and Shah [83] proposed to solve the segmentation problem by minimizing the energy:

$$E(g, \gamma) = \lambda \int_\Omega (f - g)^2 + \mu \int_{\Omega - \gamma} \|\nabla g\|^2 dx + length(\gamma)$$

The Mumford-Shah functional has had a profound impact on image processing, providing a variational framework for addressing the challenging problem of image segmentation. Its mathematical elegance and ability to capture piecewise smooth structures have made it a foundational model in the field. Ongoing research seeks to refine the model, develop efficient computational methods, and explore its applications in various domains.

Nevertheless, the research and application of the model is very difficult. In addition to the typical difficulties of regularization methods, such as the determination of regularization parameters, the treatment of the Mumford-Shah functional presents other challenges. The Mumford-Shah functional involves a non-convex optimization problem, making it challenging to find globally optimal solutions. Non-convexity leads to the presence of multiple local minima, making it difficult to ensure convergence to the desired segmentation. Solving the Mumford-Shah optimization problem often requires iterative numerical methods [66], which can be computationally intensive, especially for high-dimensional images.

The application of these ideas to computer vision problems was also proposed in the 1970s and 1980s. David Marr's research [77] on the levels of analysis to study any computational system aroused great interest in early vision issues, including the problem of visible surface reconstruction. Grimson [54, 55] characterizes this problem from the variational point of view, defining functionals on the surface to be reconstructed and proposes some interpolation methods on dispersed data currently well known as the Duchon thin plate spline [34], in which we want to find the function or surface that minimizes the functional $\mathscr{J}[u] = \int_\Omega u_{xx}^2 + 2u_{xy}^2 + u_{yy}^2$ and interpolates the scattered data $(\mathbf{x}_k, z_k)$ in $\mathbb{R}^2 \times \mathbb{R}$ such that $u(\mathbf{x}_k) = z_k$. The functional $\mathscr{J}[u]$ represents the potential energy of deformation of a thin plate. In fact, the term "energy methods" is very frequently used to refer to variational or regularization methods.

Sometimes the energy in its physical meaning, is given as a functional, and obviously, they are called **energy functionals**, for example the strain energy of an elastic beam can be approximated with the functional $\mathscr{J}[u] = \int_a^b (u'')^2 dx$; which is applied in the characterization of the well-known cubic natural spline [64, 65, 122]. During the evolution and emergence of new applications of variational calculus became a common use to call energy functionals to objective functions, although they may not have an evident physical interpretation.

Chapter 2
Mathematical Preliminaries

Euclidean geometry takes as its starting point, concepts that are considered basic or elementary and allow us to analyze the problems of space by applying our intuition. The need to extend these notions of Euclidean space to more complex objects gives rise to the general definition of a vector space. In this way we can think of these spaces with our basic geometric intuitions such as angle, distance, parallelism and orthogonality.

This chapter provides the essential mathematical background required for understanding and applying variational methods in image processing. We begin by revisiting the fundamental properties of Euclidean space and matrices, which form the building blocks for multidimensional analysis.

The concept of derivatives is then extended to multiple dimensions, including a discussion of Taylor's theorem, which serves as a foundation for local approximation of functions.

The chapter also covers key topics in optimization in $\mathbb{R}^N$, introducing techniques for solving problems involving multiple variables.

Further, we explore the structure of linear spaces, including function spaces, and delve into normed spaces, emphasizing their role in measuring distances and convergence. The chapter concludes with an introduction to inner product spaces and Hilbert spaces, which are indispensable tools for understanding variational methods and optimization in infinite-dimensional settings [89, 90].

2.1 The Space $\mathbb{R}^N$

The set of all N-dimensional points (or vectors) is called N-dimensional Euclidean space or simply N-space, and is denoted by $\mathbb{R}^N$. In the Euclidian linear space $\mathbb{R}^N$ a vector **x** is defined as a column or a $N \times 1$ matrix

$$\mathbf{x} = \begin{bmatrix} x_1 \\ x_2 \\ \vdots \\ x_N \end{bmatrix}$$

It is very common to write $\mathbf{x}^T = [x_1, x_2, \ldots, x_N]$. This definition of N-dimensional vector, not only has a geometric interpretation but many more. In physics may be a vector meaning velocity, force or acceleration, in statistics a sample vector or a collection of data coming from some variable. That $\mathbb{R}^N$ is a linear or vector space basically means:

Definition 2.1 Let $\mathbf{x} = [x_1, \ldots, x_N]^T, \mathbf{y} = [y_1, \ldots, y_N]^T$ be in $\mathbb{R}^N$. Then is defined:

1. *Equality:*

$$\mathbf{x} = \mathbf{y} \text{ if, and only, if, } x_1 = y_1, \ldots x_N = y_N$$

2. *Sum:*

$$\mathbf{x} + \mathbf{y} = [x_1 + y_1, \ldots, x_N + y_N]^T$$

3. *Scalar multiplication:*

$$\alpha \mathbf{x} = [\alpha x_1, \cdots, \alpha x_N]^T; \alpha \in \mathbb{R}$$

4. *Difference:*

$$\mathbf{x} - \mathbf{y} = \mathbf{x} + (-1)\mathbf{y}$$

5. *Zero vector:*

$$\mathbf{0} = [0, \cdots, 0]^T$$

The norm or length $\|\cdot\|$ (or $|\cdot|$) of $\mathbf{x}$ is

$$\|\mathbf{x}\| = |\mathbf{x}| := \left(\sum_{k=1}^{N} x_k^2\right)^{1/2} \tag{2.1}$$

Example If $\|\mathbf{x} - \mathbf{y}\|$ is the distance between $\mathbf{x}$ and $\mathbf{y}$, the inner product $\mathbf{x} \cdot \mathbf{y}$ has a geometrical and algebraic interpretation in $\mathbb{R}^3$ which can be seen with the Law of Cosines (Fig. 2.1).

2.1 The Space $\mathbb{R}^N$

Fig. 2.1 The idea of inner product explained from the cosine law

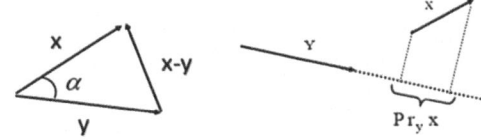

$$\mathbf{x} = \begin{bmatrix} x_1 \\ x_2 \\ x_3 \end{bmatrix} ; \mathbf{y} = \begin{bmatrix} y_1 \\ y_2 \\ y_3 \end{bmatrix}$$

$$\|\mathbf{x} - \mathbf{y}\|^2 = \|\mathbf{x}\|^2 + \|\mathbf{y}\|^2 - 2\|\mathbf{x}\|\|\mathbf{y}\|\cos\alpha$$

$$(x_1 - y_1)^2 + (x_2 - y_2)^2 + (x_3 - y_3)^2 = x_1^2 + x_2^2 + x_3^2$$
$$+ y_1^2 + y_2^2 + y_3^2 - 2\|\mathbf{x}\|\|\mathbf{y}\|\cos\alpha$$

$$\underbrace{\|\mathbf{x}\|\|\mathbf{y}\|\cos\alpha}_{\mathbf{x}\cdot\mathbf{y}} = x_1 y_1 + x_2 y_2 + x_3 y_3$$

Then we have a geometric definition and an algebraic interpretation for the inner product in $\mathbb{R}^N$ □

Definition 2.2 If α is the angle between the vectors $\mathbf{x}, \mathbf{y}$ in $\mathbb{R}^N$, their scalar or inner product is

$$\langle \mathbf{x}, \mathbf{y} \rangle = \mathbf{x} \cdot \mathbf{y} := \|\mathbf{x}\|\|\mathbf{y}\|\cos\alpha \qquad (2.2)$$

and can be calculated in matrix multiplication as

$$\langle \mathbf{x}, \mathbf{y} \rangle = \mathbf{x}^T \mathbf{y} = [x_1, x_2 \cdots, x_N] \begin{bmatrix} y_1 \\ y_2 \\ \vdots \\ y_N \end{bmatrix} = \sum_{k=1}^{N} x_k y_k \qquad (2.3)$$

then we can see

$$\|\mathbf{x}\|^2 = \mathbf{x} \cdot \mathbf{x}$$

and

$$Pr_\mathbf{y}\mathbf{x} = \frac{\mathbf{x} \cdot \mathbf{y}}{\|\mathbf{y}\|}$$

Example Unlike $\mathbf{x}^T\mathbf{y}$ which produces a real number, the operation $\mathbf{xy}^T$ is the complete opposite

$$\mathbf{xy}^T = \begin{bmatrix} x_1 \\ x_2 \\ \vdots \\ x_N \end{bmatrix} [y_1, y_2 \cdots, y_N] = \begin{bmatrix} x_1y_1 & x_1y_2 & \cdots & x_1y_N \\ x_2y_1 & x_2y_2 & \cdots & x_2y_N \\ \vdots & \vdots & \ddots & \vdots \\ x_Ny_1 & x_Ny_2 & \cdots & x_Ny_N \end{bmatrix}$$

Definition 2.3 The open ball $B_r(a)$ in $\mathbb{R}^N$ with center a and radius r is the set

$$B_r(a) = \{\mathbf{x} \in \mathbb{R}^N : \|\mathbf{x} - a\| < r\}$$

The closed ball $\bar{B}_r(a)$ in $\mathbb{R}^N$ with center a and radius r

$$\bar{B}_r(a) = \{\mathbf{x} \in \mathbb{R}^N : \|\mathbf{x} - a\| \leq r\}$$

The ball $B_r(a)$ for some arbitrary $r > 0$ is also referred to as a neighborhood of a.

2.2 Matrices $\mathbb{R}^{m \times n}$

The set $\mathbb{R}^{m \times n}$ of all real-valued matrices of order $m \times n$ is a linear space with addition $A + B$ and scalar multiplication αA.

$$A = \begin{bmatrix} a_{11} & a_{12} & \cdots & a_{1n} \\ a_{21} & a_{22} & \cdots & a_{2m} \\ \vdots & \vdots & \ddots & \vdots \\ a_{m1} & a_{m2} & \cdots & a_{mn} \end{bmatrix} \in \mathbb{R}^{m \times n}$$

Eigenvalues and Eigenvectors

Let $A \in \mathbb{R}^{n \times n}$. A nonzero vector $\mathbf{x} \in \mathbb{R}^n$ is called an **eigenvector** of A if there exists a $\lambda \in \mathbb{C}$ for which

$$A\mathbf{x} = \lambda \mathbf{x}$$

The scalar λ is the **eigenvalue** corresponding to the eigenvector $\mathbf{x}$. A real-valued matrix may have complex eigenvalues, but it is well known that all the eigenvalues

2.2 Matrices $\mathbb{R}^{m \times n}$

of a symmetric matrix are real. As a consequence we could give an order to its eigenvalues in the form $\lambda_1 \geq \lambda_2 \geq \cdots \geq \lambda_n$

Theorem 2.1 (Spectral Decomposition) *Let $A \in \mathbb{R}^{n \times n}$ be a symmetric matrix. Then there exists an orthogonal matrix $U \in \mathbb{R}^{n \times n}$ and a diagonal matrix D such that*

$$U^T A U = D \qquad (2.4)$$

The Columns of the matrix U in (2.4) form an orthonormal basis of eigenvectors for A and the diagonal of D are the corresponding eigenvalues $\{\lambda_k\}_{k=1}^n$. As an additional result we have

$$tr(A) = \sum \lambda_k$$

$$\det(A) = \prod \lambda_k$$

Singular Value Decomposition of a Matrix

According to what we have seen so far, in order to find useful properties of a matrix it is necessary that it have some previous structure or some pattern that we can take advantage of. However, there is a property that can be applied to any matrix without asking for anything in return. It is the SVD factorization.

If A is a matrix in $\mathbb{R}^{m \times N}$ then $A^T A$ it is symmetric and its eigenvalues are $\lambda_k \geq 0$. The **singular values** of A are defined as $\sigma_k = \sqrt{\lambda_k}$. The SVD (Singular Value Decomposition) express A as product of three matrices.

Theorem 2.2 (SVD) *For a matrix $A \in \mathbb{R}^{m \times N}$ with r singular values and $m \geq N$. A can be factorized as*

$$A = U \Sigma V^T \qquad (2.5)$$

in such a manner that $U \in \mathbb{R}^{m \times m}$, $V \in \mathbb{R}^{N \times N}$ y Σ is a diagonal matrix with the structure

$$\Sigma = \begin{bmatrix} \bar{\Sigma} & 0 \\ 0 & 0 \end{bmatrix} \qquad \bar{\Sigma} = \begin{bmatrix} \sigma_1 & & \\ & \ddots & \\ & & \sigma_r \end{bmatrix}$$

Here $\sigma_1 \geq \sigma_2 \geq \cdots \geq \sigma_r > 0$, are the positive singular values.

Explaining it briefly, the spectral theorem guarantees that $A^T A$ has an orthonormal basis of eigenvectors $\{\mathbf{v}_k\}_{k=1}^{N}$ of $\mathbb{R}^N$, these vectors put in column form the matrix

$$V = [\mathbf{v}_1, \mathbf{v}_2, \cdots, \mathbf{v}_N]$$

and then taking

$$\mathbf{u}_k = \frac{A\mathbf{v}_k}{\sigma_k}$$

yields the matrix $U = [\mathbf{u}_1, \mathbf{u}_2, \ldots, \mathbf{u}_m]$. Expressing the SVD factorization in matrix form

$$A = U\Sigma V^T = \begin{bmatrix} u_{11} & u_{12} & \cdots & u_{1m} \\ u_{21} & u_{22} & \cdots & u_{2m} \\ \vdots & \vdots & \ddots & \vdots \\ u_{m1} & u_{m2} & \cdots & u_{mm} \end{bmatrix} \begin{bmatrix} \sigma_1 & & & 0 \\ & \ddots & & \\ & & \sigma_r & \\ 0 & & & \ddots \\ & & & & 0 \end{bmatrix} \begin{bmatrix} v_{11} & v_{12} & \cdots & v_{1m} \\ v_{21} & v_{22} & \cdots & v_{2m} \\ \vdots & \vdots & \ddots & \vdots \\ v_{m1} & v_{m2} & \cdots & v_{mm} \end{bmatrix}$$

and computing the matrix multiplication

$$A = \sigma_1 \mathbf{u}_1 \mathbf{v}_1^T + \sigma_2 \mathbf{u}_2 \mathbf{v}_2^T + \cdots + \sigma_r \mathbf{u}_r \mathbf{v}_r^T = \sum_{k=1}^{r} \sigma_k \mathbf{u}_k \mathbf{v}_k^T \qquad (2.6)$$

Example (SVD Application) The following code implement the SVD representation (2.6) of an image and the results obtained when taking different amount of singular values σ_k. Figure 2.2 shows the result of representing the image including only the first 25 singular values. We can see that these representation is enough to preserve a great amount of the original information (Fig. 2.3).

Fig. 2.2 SVD for an image compression. svdtest2.m

2.3 Derivatives

Fig. 2.3 Graph of singular values for the image of Lena in decreasing order

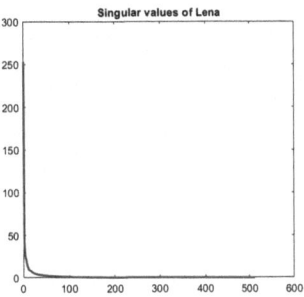

Listing 2.1 SVD

```
function sigma=svdtest2(k)
I=imread('lena.bmp');
subplot(1,2,1), imshow(I)
I = im2double(I);
[U,S,V] = svd(I);
sigma = diag(S);
p=length(sigma)
I1=0;
for i=1:k
I1 = I1 + sigma(i)*U(:,i)*V(:,i)';
end
sigma(1:k);
subplot(1,2,2), imshow(I1)
```

2.3 Derivatives

In the classical formulation (1.6) we can observe that obtaining the Lagrange equations consists of formulating the problem in terms of the derivative of a function $\Phi(\varepsilon)$ in one variable. In fact, the definition of derivative of univariate functions and all the consequences that follow from it can be generalized and applied to variational optimization problems [5, 6, 40].

Definition 2.4 Let $f : \mathbb{R} \to \mathbb{R}$ be defined on an open interval $]a, b[$ and assume that $x \in]a, b[$. Then f is said to be differentiable at x whenever the limit

$$f'(x) := \lim_{h \to 0} \frac{f(x+h) - f(x)}{h}$$

exists. $f'(x)$ is called the derivative of f at x and is extended to multivariate functions in a natural way.

Definition 2.5 (Directional Derivative) Given $f : \Omega \subset \mathbb{R}^N \to \mathbb{R}$ and $\mathbf{x}$ an interior point of Ω and let $\mathbf{y}$ an arbitrary point in $\mathbb{R}^N$. The directional derivative of f at $\mathbf{x}$ with respect to $\mathbf{y}$ is denoted by the symbol $f'(\mathbf{x}; \mathbf{y})$ and is defined by

$$f'(\mathbf{x}; \mathbf{y}) = \lim_{t \to 0} \frac{f(\mathbf{x} + t\mathbf{y}) - f(\mathbf{x})}{t} \qquad (2.7)$$

when the limit exists. The following result can be seen as a way to calculate $f'(\mathbf{x}; \mathbf{y})$

Theorem 2.3 *If we consider*

$$g(t) := f(\mathbf{x} + t\mathbf{y}) \qquad (2.8)$$

then

$$g'(t) = f'(\mathbf{x} + t\mathbf{y}; \mathbf{y}) \qquad (2.9)$$

in particular

$$f'(\mathbf{x}; \mathbf{y}) = g'(0) = \frac{\partial}{\partial t} f(\mathbf{x} + t\mathbf{y}) \Big|_{t=0} \qquad (2.10)$$

This fact is easily seen with the equality

$$\frac{g(t+h) - g(t)}{h} = \frac{f(\mathbf{x} + t\mathbf{y} + h\mathbf{y}) - f(\mathbf{x} + t\mathbf{y})}{h}$$

Example Compute $f'(\boldsymbol{a}; \mathbf{y})$ for $f(\mathbf{x}) = \|\mathbf{x}\|^2$.

Solution $f(\boldsymbol{a} + t\mathbf{y}) = (\boldsymbol{a} + t\mathbf{y}) \cdot (\boldsymbol{a} + t\mathbf{y}) = \boldsymbol{a} \cdot \boldsymbol{a} + 2t\boldsymbol{a} \cdot \mathbf{y} + t^2 \mathbf{y} \cdot \mathbf{y}$. Then

$$\frac{\partial}{\partial t} f(\boldsymbol{a} + t\mathbf{y}) = 2\boldsymbol{a} \cdot \mathbf{y} + 2t\mathbf{y} \cdot \mathbf{y}$$

and by (2.10)

$$f'(\boldsymbol{a}; \mathbf{y}) = \frac{\partial}{\partial t} f(\boldsymbol{a} + t\mathbf{y}) \Big|_{t=0} = 2\boldsymbol{a} \cdot \mathbf{y}$$

□

Definition 2.6 The basis vectors $\boldsymbol{e}_k$ in $\mathbb{R}^N$ are the vectors with 1 in the k-th component and 0 in the remaining components

$$\boldsymbol{e}_1 = [1, 0, 0, \ldots 0]^T, \boldsymbol{e}_2 = [0, 1, \ldots 0]^T, \ldots, \boldsymbol{e}_N = [0, 0, \ldots, 1]^T$$

then the partial derivatives $\frac{\partial f}{\partial x_k}$ are defined as

$$\frac{\partial f}{\partial x_k}(\mathbf{x}) := f'(\mathbf{x}; \boldsymbol{e}_k)$$

and the **Gradient** $\nabla f(\mathbf{x})$ (or $\frac{\partial f}{\partial \mathbf{x}}$ when f is written in terms of $\mathbf{x}$) of f at $\mathbf{x}$ is defined as the vector

2.3 Derivatives

$$\nabla f(\mathbf{x}) := \left[\frac{\partial f}{\partial x_1}, \ldots, \frac{\partial f}{\partial x_N}\right]^T$$

Example

(a) If $\mathbf{w}(t) = \mathbf{x} + t\mathbf{y}$ then $g(t) = f(\mathbf{x} + t\mathbf{y}) = f(\mathbf{w}(t))$, and by the chain rule

$$\frac{\partial}{\partial t} f(\mathbf{x} + t\mathbf{y}) = \sum \frac{\partial f}{\partial w_i} \frac{dw_i}{dt} = \sum \frac{\partial f}{\partial w_i} y_i$$

$$= \nabla f(\mathbf{x} + t\mathbf{y})^T \mathbf{y}$$

and by (2.10)

$$\frac{\partial}{\partial t} f(\mathbf{x} + t\mathbf{y})|_{t=0} = \nabla f(\mathbf{x})^T \mathbf{y} = \langle \nabla f(\mathbf{x}), \mathbf{y} \rangle \qquad (2.11)$$

(b) By the fundamental theorem of calculus

$$g(1) = g(0) + \int_0^1 g'(t) dt$$

and replacing in the definition of $g(t)$ and (2.11)

$$f(\mathbf{x} + \mathbf{y}) = f(\mathbf{x}) + \int_0^1 \nabla f(\mathbf{x} + t\mathbf{y})^T \mathbf{y} \, dt \qquad (2.12)$$

(c) By the mean value theorem for one variable functions, there exists ξ in $]0, 1[$ such that

$$g(1) = g(0) + g'(\xi)$$

then by (2.11) and (2.8)

$$f(\mathbf{x} + \mathbf{y}) = f(\mathbf{x}) + \nabla f(\mathbf{x} + \xi \mathbf{y})^T \mathbf{y}$$

for all ξ in $]0, 1[$. □

Example Calculate the directional derivative for $f(\mathbf{x}) = \mathbf{x}^T A \mathbf{x}$, with $A \in \mathbb{R}^{N \times N}$, $\mathbf{x} \in \mathbb{R}^N$

$$f(\mathbf{x} + t\mathbf{y}) = (\mathbf{x} + t\mathbf{y})^T A (\mathbf{x} + t\mathbf{y})$$
$$= \mathbf{x}^T A \mathbf{x} + t\mathbf{x}^T A \mathbf{y} + t\mathbf{y}^T A \mathbf{x} + t^2 \mathbf{y}^T A \mathbf{y}$$

$$\frac{\partial}{\partial t} f(\mathbf{x} + t\mathbf{y}) = \mathbf{x}^T A \mathbf{y} + \mathbf{y}^T A \mathbf{x} + 2t \mathbf{y}^T A \mathbf{y}$$

then

$$f'(\mathbf{x}; \mathbf{y}) = \frac{\partial}{\partial t} f(\mathbf{x} + t\mathbf{y})\Big|_{t=0} = (A^T\mathbf{x} + A\mathbf{x})^T \mathbf{y}$$

and using (2.11) we can say $\nabla f = A^T\mathbf{x} + A\mathbf{x} = (A^T + A)\mathbf{x}$. In case A symmetric we obtain $\nabla f = 2A\mathbf{x}$ which is a well-known formula. □

Example (Gradient of a Quadratic Form) A quadratic function over $\mathbb{R}^N$ is a function of the form

$$f(\mathbf{x}) = \mathbf{x}^T A\mathbf{x} + 2\mathbf{b}^T \mathbf{x} + c$$

with $\mathbf{x}^T A\mathbf{x} = \sum_{i=1}^{N} \sum_{j=1}^{N} a_{ij} x_i x_j$. Then applying the former results, its gradient is,

$$\frac{\partial f}{\partial \mathbf{x}} = \nabla f(\mathbf{x}) = 2A\mathbf{x} + 2\mathbf{b} \qquad (2.13)$$

Example (A Least Square Problem) Given the data

$$(x_1, y_1), (x_2, y_2) \cdots, (x_N, y_N)$$

in $\mathbb{R}^2$, find the line $y = f(x) = mx + b$ that minimize the least square error functional

$$E(\mathbf{x}) = \sum_{k=1}^{N} (y_k - f(x_k))^2 = \|\mathbf{y} - A\mathbf{x}\|^2$$

$$y = \begin{bmatrix} y_1 \\ y_2 \\ \vdots \\ y_N \end{bmatrix}, \quad A = \begin{bmatrix} 1 & x_1 \\ 1 & x_2 \\ \vdots & \vdots \\ 1 & x_N \end{bmatrix} \quad \text{and } \mathbf{x} = \begin{bmatrix} b \\ m \end{bmatrix}$$

Then we have to minimize the functional

$$E(\mathbf{x}) = \|A\mathbf{x} - \mathbf{y}\|^2 = \langle A\mathbf{x} - \mathbf{y}, A\mathbf{x} - \mathbf{y}\rangle = \mathbf{x}^T(A^T A)\mathbf{x} - 2(A^T \mathbf{y})^T \mathbf{x} + \|\mathbf{y}\|^2$$

applying (2.13) we have

$$\nabla E = 2A^T A\mathbf{x} - 2A^T \mathbf{y} = 0$$

obtaining the so called **normal equations**

$$A^T A\mathbf{x} = A^T \mathbf{y} \qquad (2.14)$$

Fig. 2.4 Edge detection in the image f by values of $\nabla f(x, y)$

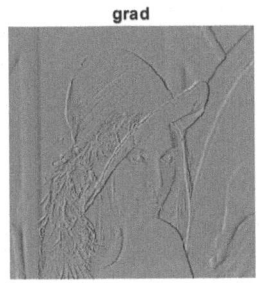 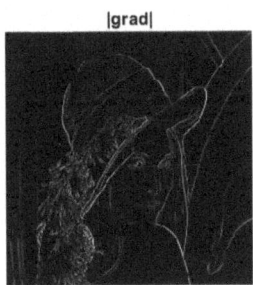

and $\mathbf{x} = (A^T A)^{-1} A^T \mathbf{y}$. □

Example (Divergence and Laplacian)

- Given a vector field $\mathcal{F} = (f_1, f_2, \ldots, f_N)$ the divergence $div\mathcal{F} \equiv \nabla \cdot \mathcal{F}$ of $\mathcal{F}$ is

$$\nabla \cdot \mathcal{F} := \sum_{k=1}^{N} \frac{\partial f_k}{\partial x_k}$$

- Given $f : \mathbb{R}^N \to \mathbb{R}$ The Laplacian Δf of f is

$$\Delta f := div \circ \nabla = \sum_{k=1}^{N} \frac{\partial^2 f}{\partial x_k^2}$$

Example (Derivatives and Edge Detection) Intensity discontinuities in an image can be used for defining boundaries of objects and regions. This is part of the problems of edge detection. These sharp changes in image brightness can be detected using derivatives $f'(x)$ for one-dimensional signals and $\nabla f(x, y)$ for images. In the Fig. 2.4 we can see an eloquent example of how the gradient values can detect edges in an image. □

2.4 Taylor's Theorem

Theorem 2.4 *If $f \in C^n[a, b]$ and if $f^{(n+1)}$ exists on the open interval $]a, b[$ then for any points x_0 and x in the interval $[a, b]$,*

$$f(x) = \sum_{k=0}^{N} \frac{f^{(k)}(x_0)}{k!} (x - x_0)^k + R_N(x)$$

where, for some point ξ between x_0 and x, the error term is

$$R_N(x) = \frac{f^{(N+1)}(\xi)}{(N+1)!}(x-x_0)^{N+1}$$

For a scalar function $f : \mathbb{R}^2 \to \mathbb{R}$ with $\mathbf{x}_0 = (x, y)$; $\mathbf{x} = (h, k)$ Taylor series holds in the form

$$f(x+h, y+k) = \sum_{p=0}^{\infty} \frac{1}{p!}\left(h\frac{\partial}{\partial x} + k\frac{\partial}{\partial y}\right)^p f(\mathbf{x}_0) \qquad (2.15)$$

$$= f(\mathbf{x}_0) + h\frac{\partial f(\mathbf{x}_0)}{\partial x} + k\frac{\partial f(\mathbf{x}_0)}{\partial y}$$

$$+ \frac{1}{2!}\left(h^2\frac{\partial^2 f(\mathbf{x}_0)}{\partial x^2} + 2hk\frac{\partial^2 f(\mathbf{x}_0)}{\partial x \partial y} + k^2\frac{\partial^2 f(\mathbf{x}_0)}{\partial y^2}\right) + \cdots$$

$$\qquad (2.16)$$

$$= f(\mathbf{x}) + \underbrace{\begin{bmatrix} \frac{\partial f}{\partial x} & \frac{\partial f}{\partial y} \end{bmatrix}}_{\nabla f^T}\begin{bmatrix} h \\ k \end{bmatrix} + \frac{1}{2!}[h\ k]\underbrace{\begin{bmatrix} \frac{\partial^2}{\partial x^2} & \frac{\partial^2}{\partial x \partial y} \\ \frac{\partial^2}{\partial y \partial x} & \frac{\partial^2}{\partial y^2} \end{bmatrix}}_{\nabla^2 f}\begin{bmatrix} h \\ k \end{bmatrix} + \cdots$$

$$\qquad (2.17)$$

The first two terms in (2.15) form the tangent plane. In general, for a function $f : \mathbb{R}^N \to \mathbb{R}$ we have

Theorem 2.5 (Taylor's) *Suppose that* $f : \mathbb{R}^N \to \mathbb{R}$ *is twice continuously differentiable and* $\mathbf{x} \in \mathbb{R}^N$ *then*

$$f(\mathbf{x}_0 + \mathbf{x}) = f(\mathbf{x}_0) + \nabla f(\mathbf{x}_0)^T \mathbf{x} + \frac{1}{2!}\mathbf{x}^T \nabla^2 f(\mathbf{x}_0 + \xi \mathbf{x})\mathbf{x}, \qquad \xi \in]0, 1[\qquad (2.18)$$

This can also be written as

$$f(\mathbf{x}_0 + \mathbf{x}) = f(\mathbf{x}_0) + \nabla f(\mathbf{x}_0)^T \mathbf{x} + \frac{1}{2!}\mathbf{x}^T \nabla^2 f(\mathbf{x}_0)\mathbf{x} + \|\mathbf{x}\|^2 E_2(\mathbf{x}_0, \mathbf{x}), \qquad (2.19)$$

$E_2(\mathbf{x}_0, \mathbf{x}) \to 0$ as $\mathbf{x} \to 0$. $\nabla^2 f(\mathbf{x}_0)$ *is called the* **Hessian matrix**.

2.5 Optimization in $\mathbb{R}^N$

Definition 2.7 A function $f : \mathbb{R}^N \to \mathbb{R}$ is said to have an absolute minimum at a point $\mathbf{x}_0 \in \Omega \subset \mathbb{R}^N$ if

$$f(\mathbf{x}_0) \leq f(\mathbf{x}) \qquad (2.20)$$

2.5 Optimization in $\mathbb{R}^N$

Fig. 2.5 Geometric interpretation of ∇f as normal vector to the surface $f(\mathbf{x}) = c$

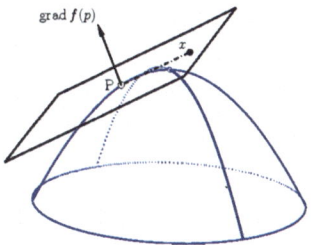

for all $\mathbf{x}$ in Ω. The number $f(\mathbf{x}_0)$ is called the **global minimum** value of f on Ω. The function f is said to have a **local minimum** at $\mathbf{x}_0$ if (2.20) is satisfied for every $\mathbf{x}$ in some N-ball $B(\mathbf{x}_0)$ in Ω.

In other words, a local minimum at $\mathbf{x}_0$ is the global minimum in some neighborhood of $\mathbf{x}_0$. Using the opposite inequality in (2.20) can be defined the terms global maximum and local maximum. The terms absolute and relative are also used for global and local, respectively. A number which is either a local maximum or a local minimum of f is called an **extremum** of f [5, 6].

It is well known that if $f(x, y, z) = c$ is a surface Σ in $\mathbb{R}^3$ then ∇f is normal to the surface. This can be seen considering a curve $r(t)$ on Σ, then $f(r(t)) = c$ and by the chain rule $\nabla f \perp r'(t)$ (Fig. 2.5). This geometric interpretation suggests that if f has an extremum at an interior point $\mathbf{x}_0$ and is differentiable there, then all first-order partial derivatives $\frac{\partial f}{\partial x_k}$ must be zero, this is $\nabla f = 0$

Theorem 2.6 (First Order Necessary Conditions) *Let $f : \Omega \subset \mathbb{R}^N \to \mathbb{R}$ be a function and $\mathbf{x}_0$ an interior point of Ω. If $\mathbf{x}_0$ is an extremum of f and all the partial derivatives of f exist at $\mathbf{x}_0$ then $\nabla f(\mathbf{x}_0) = 0$*

Proof For $k \in \{1, 2, \ldots, N\}$ define the function $g(t) := f(\mathbf{x}_0 + t\mathbf{e}_k)$. Since $\mathbf{x}_0$ is an extremum of f then $t = 0$ is an extremum of g it follows that $g'(0) = \frac{\partial f}{\partial x_k} = 0$ for $k = 1, 2, \ldots, N$ then $\nabla f(\mathbf{x}_0) = 0$ □

Definition 2.8 (Descent Direction) If $f : \mathbb{R}^N \to \mathbb{R}$ is a continuous differentiable function and its directional derivative $f'(\mathbf{x}_0; \mathbf{v}) = \nabla f(\mathbf{x}_0)^T \mathbf{v} < 0$; then $\mathbf{v} \in \mathbb{R}^N$; $\mathbf{v} \neq 0$ is called a descent direction of f at $\mathbf{x}_0$.

The most important property of descent directions is their applications in finding stationary points for a function. Going along these directions lead to decreasing values of the function f.

If $f : \mathbb{R}^N \to \mathbb{R}$ is a continuous differentiable function on $\mathbb{R}^N$ and $\mathbf{x} \in \mathbb{R}^N$, if $\mathbf{v}$ is a descent director of f at $\mathbf{x}$. Then there exists $\varepsilon > 0$ such that

$$f(\mathbf{x} + t\mathbf{v}) < f(\mathbf{x}); \qquad \forall t \in]0, \varepsilon]$$

An obvious way to follow this result is to create iterative algorithms such that the stationary point is approximated by

$$\mathbf{x}_{k+1} = \mathbf{x}_k + t_k \mathbf{v}_k; \qquad k = 0, 1, 2, ..$$

In the gradient method the descent direction is chosen to be the minus of the gradient at the current point: $\mathbf{v}_k = -\nabla f(\mathbf{x}_k)$. As can be seen this is a descent direction

$$f'(\mathbf{x}_k; -\nabla f(\mathbf{x}_k)) = -\nabla f(\mathbf{x}_k)^T \nabla f(\mathbf{x}_k) = -\|\nabla f(\mathbf{x}_k)\|^2 < 0$$

Algorithm 1 Gradient descent

1: Input: $\varepsilon > 0$
2: Pick $\mathbf{x}_0 \in \mathbb{R}^N$ arbitrarily.
3: **for** $k = 1, 2, \ldots$ **do**
4:     (a) Choose $t_k \in \mathbb{R}$ by line search on

$$g(t) = f(\mathbf{x}_k - t \nabla f(\mathbf{x}_k)).$$

5:     (b) $\mathbf{x}_{k+1} = \mathbf{x}_k - t_k \nabla f(\mathbf{x}_k)$
6:     (c) If $\|\nabla f(\mathbf{x}_{k+1})\| \leq \varepsilon$ then STOP and $\mathbf{x}_{k+1}$ is the output
7: **end for**
8: Output: $\mathbf{x}_{k+1}$

Newton's Method for Optimization

Newton's method uses the quadratic approximation by Taylor series (2.19), we have

$$f(\mathbf{x}_k + \mathbf{x}) \approx f(\mathbf{x}_k) + \nabla f(\mathbf{x}_k)^T \mathbf{x} + \frac{1}{2} \mathbf{x}^T \nabla^2 f(\mathbf{x}_k) \mathbf{x} =: M_k(\mathbf{x})$$

A necessary condition or a minimizer of M_k is $\nabla M_k(\mathbf{x}) = 0$. Applying (2.13), the gradient of the quadratic form is

$$\nabla M_k = \nabla^2 f_k \mathbf{x} + \nabla f_k = 0$$

and solving for $\mathbf{x}$

$$\mathbf{x} = -(\nabla^2 f_k)^{-1} \nabla f_k$$

so the iteration of Newton's method is

$$\mathbf{x}_{k+1} = \mathbf{x}_k - (\nabla^2 f(\mathbf{x}_k))^{-1} \nabla f(\mathbf{x}_k) \qquad (2.21)$$

Algorithm 2 Newton's method

1: Input: $\varepsilon > 0$- tolerance parameter
2: Initialization: Choose $\mathbf{x}_0 \in \mathbb{R}^N$ arbitrarily.
3: **for** $k = 1, 2, \ldots$ **do**
4:     (a) Compute $\mathbf{y}_k \in \mathbb{R}^N$ as solution of the linear system

$$\nabla^2 f(\mathbf{x}_k)\mathbf{y}_k = -\nabla f(\mathbf{x}_k)$$

5:     (b) $\mathbf{x}_{k+1} = \mathbf{x}_k + \mathbf{y}_k$
6:     (c) If $\|\nabla f(\mathbf{x}_{k+1})\| \leq \varepsilon$, then STOP, and $\mathbf{x}_{k+1}$ is the output
7: **end for**
8: Output: $\mathbf{x}_{k+1}$

2.6 Some Integral Properties

Some properties of integrals, such as integration by parts, are a fundamental element in the operation of variational calculus. These properties are simplified when compactly supported functions are included, thanks to their particular property of vanishing at the boundary of the domain of integration. In the following results we assume that the functions involved are sufficiently differentiable.

$\Omega \subset \mathbb{R}^N$ is assumed to be open and bounded with a smooth boundary $\partial\Omega$. A set $\Omega \subset \mathbb{R}^N$ has a smooth boundary if at each point $\mathbf{x} \in \partial\Omega$ we can make an orthogonal change of coordinates so that for some $r > 0$, $\partial\Omega \cap B_r(0)$ is the graph of a smooth function $u : \mathbb{R}^{N-1} \to \mathbb{R}$. If $\partial\Omega$ is smooth, we can define an outward normal vector $n = n(\mathbf{x})$ at each point $\mathbf{x} \in \partial\Omega$, and ν varies smoothly with $\mathbf{x}$. Here, $n = (\nu_1, \ldots, \nu_N) \in \mathbb{R}^N$ and ν is a unit vector so $\|n\| = 1$. The normal derivative of $u \in C^1(\bar{\Omega})$ at $\mathbf{x} \in \partial\Omega$ is

$$\frac{\partial u}{\partial n} := \nabla u(\mathbf{x}) \cdot n(\mathbf{x})$$

Theorem 2.7 (Integration by Parts) *Let u and v be differentiable functions on $\Omega \cup \partial\Omega$. Then*

$$\int_\Omega \frac{\partial u}{\partial x_i} v \, dx = -\int_\Omega u \frac{\partial v}{\partial x_i} dx + \int_{\partial\Omega} u v \nu_i \, dS \qquad (2.22)$$

dS is the boundary element and ν_i the i-th component of the outer normal of Ω

In the on-dimensional case this is equivalent to well-known integration by parts formula

$$\int_a^b u' v \, dx = -\int_a^b u v' \, dx + u(b)v(b) - u(a)v(a)$$

Theorem 2.8 (Green's Formulas)

1.
$$\int_\Omega \Delta u\, dx = \int_{\partial\Omega} \frac{\partial u}{\partial n} dS$$

2.
$$\int_\Omega \nabla \cdot \nabla v\, dx = -\int_\Omega u\nabla v\, dx + \int_{\partial\Omega} u\frac{\partial v}{\partial n} dS$$

3.
$$\int_\Omega u\nabla v - v\nabla u\, dx = \int_{\partial\Omega} u\frac{\partial v}{\partial n} - v\frac{\partial u}{\partial n} dS$$

Theorem 2.9 (Divergence Theorem) *Let $\mathcal{F} = [f_1, \ldots, f_n]^T$ be a vector field defined on $\Omega \cup \partial\Omega$ and*

$$div\mathcal{F} = \nabla \cdot \mathcal{F} = \sum_{i=1}^{n} \frac{\partial f_i}{\partial x_i}$$

then

$$\int_{\partial\Omega} \mathcal{F} \cdot n\, dS = \int_\Omega \nabla \cdot \mathcal{F}\, dx$$

This is a generalization of Calculus' Fundamental theorem

$$f(b) - f(a) = \int_a^b f'(x)\, dx$$

Example Other version of divergence theorem is to consider $\varphi : \mathbb{R}^N \to \mathbb{R}$ and $\mathcal{F} : \mathbb{R}^N \to \mathbb{R}^N$. In this case the following formula holds

$$\nabla \cdot (\varphi \mathcal{F}) = \varphi \nabla \cdot \mathcal{F} + \nabla\varphi \cdot \mathcal{F}$$

It is easy to see the case for $N = 2$ and $\mathcal{F} = \{p(x, y), q(x, y)\}$.

$$\nabla \cdot (\varphi\{p, q\}) = \frac{\partial}{\partial x}(\varphi p) + \frac{\partial}{\partial y}(\varphi q)$$
$$= \varphi(p_x + q_y) + (\varphi_x p + \varphi_y q)$$
$$= \varphi \nabla \cdot \mathcal{F} + \nabla\varphi \cdot \mathcal{F}$$

2.7 Linear Spaces

Definition 2.9 A linear space, or vector space, over the field $\mathbb{R}$ of real numbers is a set $\mathcal{X}$, of elements called points, or vectors, endowed with the operations of addition and scalar multiplication having the following properties:

$\forall \mathbf{x}, \mathbf{y} \in \mathcal{X}; \forall \alpha, \beta \in \mathbb{R}$ $(\mathcal{X}, +)$ commutative group : $\forall \mathbf{x}, \mathbf{y}, \mathbf{z} \in \mathcal{X}$

$\mathbf{x} + \mathbf{y} \in \mathcal{X},$ $0 + \mathbf{x} = \mathbf{x}, \quad \exists 0 \in \mathcal{X}$

$\alpha \mathbf{x} \in \mathcal{X},$ $\mathbf{x} + (-\mathbf{x}) = 0, \quad \exists (-\mathbf{x}) \in \mathcal{X}$

$1\mathbf{x} = \mathbf{x},$ $\mathbf{x} + \mathbf{y} = \mathbf{y} + \mathbf{x},$

$\alpha(\beta \mathbf{x}) = (\alpha \beta) \mathbf{x},$ $\mathbf{x} + (\mathbf{y} + \mathbf{z}) = (\mathbf{x} + \mathbf{y}) + \mathbf{z}$

$(\alpha + \beta)\mathbf{x} = \alpha \mathbf{x} + \beta \mathbf{x},$

$\alpha(\mathbf{x} + \mathbf{y}) = \alpha \mathbf{x} + \alpha \mathbf{y}$

Example The set $\mathbb{R}$ of the real numbers is a real linear space when the addition and scalar multiplication are the usual addition and multiplication. Similarly, the set $\mathbb{C}$ of the complex numbers is a complex linear space.

In general the Euclidian Space $\mathbb{R}^N$ as defined in Sect. 2.1 is a linear vector space.

□

2.8 Function Spaces

A function space [56, 125] is a collection of functions having a common domain Ω. It is usually assumed that a function space is endowed with some sort of algebraic or topological structure. Most of the vector spaces that we will use are function spaces and in them the operations of addition and multiplication by a scalar are defined pointwise. To explain, let $\Omega \subset \mathbb{R}^N$ be any nonempty set and let $\mathcal{U}$ denote the collection of all real valued functions defined on Ω.

If $u, v \in \mathcal{U}$, then we define $u + v$ by

$$(u + v)(\mathbf{x}) = u(\mathbf{x}) + v(\mathbf{x})$$

for all $\mathbf{x} \in \Omega$ which is an element of $\mathcal{U}$ and

$$(\alpha u)(\mathbf{x}) = \alpha u(\mathbf{x})$$

which is also in $\mathcal{U}$. With these two operations $\mathcal{U}$ becomes a linear space.

Definition 2.10 A subspace $\mathcal{V}$ of the linear space $\mathcal{U}$ is a subset of $\mathcal{U}$ which is closed under the addition and scalar multiplication operations of $\mathcal{U}$, i.e., for any $u, v \in \mathcal{V}$ and any $\alpha \in \mathbb{R}$, we have $u + v \in \mathcal{V}$ and $\alpha v \in \mathcal{V}$.

Example If Ω is an open subset of $\mathbb{R}^N$ is usual to consider the following function subspaces from Ω into $\mathbb{R}$.

$C(\Omega)$ = the space of all continuous real valued functions defined on Ω.
$C^k(\Omega)$ = the space of all continuous real valued functions defined on Ω with continuous partial derivatives of order k.
$C^\infty(\Omega)$ = the space of infinitely differentiable functions defined on Ω.
$\Pi(\Omega)$ = the space of polinomials of N variables. □

If $\mathcal{V}_1, \mathcal{V}_2$ are subspaces of a vector space $\mathcal{U}$ and $\mathcal{V}_1 \subset \mathcal{V}_2$, then $\mathcal{V}_1$ is a subspace of $\mathcal{V}_2$. For example, $\Pi(\mathbb{R}^N)$ is a subspace of $C^\infty(\mathbb{R}^N)$, which in turn is a subspace of $C^k(\mathbb{R}^N)$.

$\mathcal{V}_1 \cap \mathcal{V}_2$ is always a subspace but is not true for $\mathcal{V}_1 \cup \mathcal{V}_2$.

Given vectors $u_1, u_2, \cdots, u_n \in \mathcal{U}$ and scalars $\alpha_1, \cdots, \alpha_n \in \mathbb{R}$ we call

$$\sum_{i=1}^n \alpha_i u_i = \alpha_1 u_1 + \cdots + \alpha_n u_n$$

a **linear combination** of $u_1, u_2, \cdots, u_n$.

It is logical to think that in a set of vectors there may be redundant information, so they could be excluded from the linear combination. This gives rise to the idea of linear dependence and independence.

Definition 2.11 We say $u_1, u_2, \cdots, u_m \in \mathcal{U}$ are linearly dependent if there are scalars $\alpha_i \in \mathbb{R}$, $i = 1, \cdots, m$, with at least one $\alpha_i \neq 0$ such that

$$\sum_{i=1}^m \alpha_i u_i = 0 \tag{2.23}$$

We say $u_1, u_2, \cdots, u_m$ are linearly independent if they are not linearly dependent, in other words, if (2.23) implies $\alpha_i = 0$ for $i = 1, 2, \ldots, m$.

Example In the space $C[0, 1]$, the vectors $\{1, t, t^2, \cdots, t^m\}$ are linearly independent. If we assume

$$S(t) = \sum_{j=0}^m \alpha_j t^j = 0$$

one way to show linear independence is taking the derivatives of $S(t)$ and doing $S^k(0) = 0$ then $\alpha_k = 0, k = 0, \ldots m$

Definition 2.12 The subspace

$$\text{span}\{u_1, \cdots, u_m\} := \left\{ \sum_{i=1}^m \alpha_i u_i : \alpha_i \in \mathbb{R} \right\}$$

of linear combinations, is called the span of $\{u_1, \cdots, u_m\}$.

A set of vectors $\mathcal{B} \subset \mathcal{U}$ is called a basis of $\mathcal{U}$ if $\mathcal{B}$ is linearly independent and $\text{span}\mathcal{B} = \mathcal{U}$.

If there exists a finite basis in $\mathcal{U}$, then $\mathcal{U}$ is called a finite dimensional vector space. Otherwise we say that $\mathcal{U}$ is infinite dimensional. It can be proved that, for a given finite dimensional space, the number of vectors in any of its basis is the same. For example, if $\mathcal{U}$ has a basis with n vectors, then any other basis has exactly n vectors. In such a case n is called the dimension of $\mathcal{U}$ and we write $dim\mathcal{U} = n$.

Convexity

Definition 2.13 A set C in a linear vector space is said to be **convex** if, given $\mathbf{x}_1, \mathbf{x}_2 \in C$, all points of the form $\alpha \mathbf{x}_1 + (1 - \alpha)\mathbf{x}_2$ such that $0 < \alpha < 1$, are in C.

In other words, if we consider two points in a convex set, the line segment between them is also in the set. The empty set is convex.

Some properties are

- If C is convex then $\alpha C = \{\mathbf{x} : \mathbf{x} = \alpha \mathbf{w}, \mathbf{w} \in C\}$ is convex
- $C1, C2$ convex sets, then $C1 + C_2$ convex
- The set $\bigcap_{C\, convex} C$, is convex

2.9 Analysis on Normed Spaces

Definition 2.14 Let $\mathcal{U}$ be a vector space. A function

$$\|\cdot\| : \mathcal{U} \to [0, \infty]$$

is called a **norm** if it has the following properties:

1. $\|u\| \geq 0$, and $\|u\| = 0$ if, and only if, $u = 0$
2. $\|\alpha u\| = |\alpha| \|u\|$; for every real α
3. $\|u + v\| \leq \|u\| + \|v\|$

The pair $(\mathcal{U}, \|\cdot\|)$ is called a **normed vector space**

Example For $\mathbf{x} = [x_1, \cdots, x_N]^T$ the formula

$$\|\mathbf{x}\|_2 := \left(\sum_{k=1}^{N} x_k^2\right)^{1/2}$$

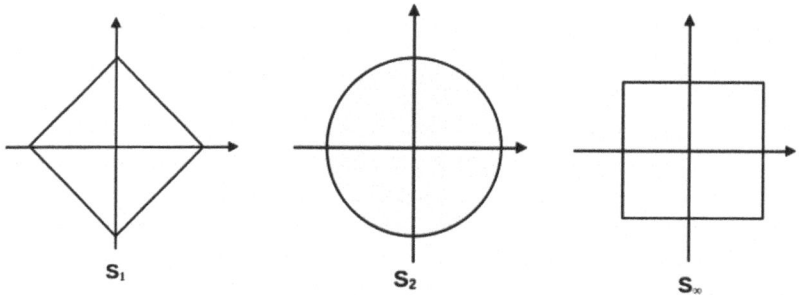

Fig. 2.6 The unit ball $S_p = \{\mathbf{x} \in \mathbb{R}^2 \|\mathbf{x}\|_p = 1\}$ for $p = 1, 2, \infty$

defines the usual Euclidian norm in $\mathbb{R}^N$. If $N = 1$, the norm is equivalent to the absolute value: $\|x\| = |x|$ for $x \in \mathbb{R}$. The subscript 2 will be frequently omitted. This is a particular case of the p-**norm**

$$\|\mathbf{x}\|_p := \left(\sum_{k=1}^{N} |x_k|^p\right)^{1/p}$$

as well as the l_1 norm

$$\|\mathbf{x}\|_1 = \sum_{k=1}^{N} |x_k|$$

and the maximum or infinity norm

$$\|\mathbf{x}\|_\infty := \max_{k=1,2,\cdots,N} |x_k|$$

such that $\|\mathbf{x}\|_\infty = \lim_{p\to\infty} \|\mathbf{x}\|_p$ (Fig. 2.6) □

Example The standard norm in $C[a, b]$ is the maximum norm

$$\|u\|_\infty = \max_{a \leq x \leq b} |u(x)|, u \in C[a, b]$$

Example Let f be a real-valued function on an interval $[a, b] \subset \mathbb{R}$ and a partition $P = \{x_k\}_{k=0}^{n}$ on $[a, b]$ such that $a = x_0 < x_1 < x_2 < \cdots < x_n = b$. The variation of the function f over a partition P is defined to be

$$V(f; P) := \sum_{i=0}^{n-1} |f(x_{i+1}) - f(x_i)|. \tag{2.24}$$

2.9 Analysis on Normed Spaces

The Total Variation of a function is defined to be the supremum of the variation over all possible partitions. Suppose $\mathcal{P}$ is the collection of all partitions defined on the interval $[a, b]$. Then the Total Variation of f is given by

$$V(f) := \sup_{P \in \mathcal{P}} \sum_{i=0}^{n-1} |f(x_{i+1}) - f(x_i)| \tag{2.25}$$

A function is defined to be a function of bounded variations provided $V(f) < \infty$. The set of all functions of bounded variation defined on the interval $[a, b]$ is denoted $BV[a, b]$. The variation of a function $V(f)$ is equal to zero if and only if the function f is identically equal to a constant on $[a, b]$. Thus, $V(\cdot)$ is not a norm on $BV[a, b]$. It is simple to modify it, however, to define a norm on $BV[a, b]$. For example, the expression

$$\|f\| := V(f) + |f(a)|$$

defines a norm on $BV[a, b]$. To show that $\|\cdot\|$ defines a norm on $BV[a, b]$, we have $\|\alpha f\| = V(\alpha f) + |\alpha f(a)| = |\alpha| V(f) + |\alpha| |f(a)| = |\alpha| \|f\|$ and satisfies a triangle inequality. By definition $\|f + g\| = V(f + g) + |f(a) + g(a)|$. Then

$$V(f+g) = \sup_{P \in \mathcal{P}} \sum_i |f_{i+1} + g_{i+1} - f_i - g_i|$$

$$\leq \sup_{P \in \mathcal{P}} \sum_i |f_{i+1} - f_i| + |g_{i+1} - g_i|$$

$$\leq V(f) + V(g)$$

and the triangle inequality holds. □

Definition 2.15 (Open and Closed Balls) If u is an element in a normed space $\mathcal{U}$, and r a positive number, we use the same notation for open and closed balls given in Definition 2.3

Definition 2.16 (Open and Closed Sets) A subset S of a normed space $\mathcal{U}$ is called **open** if for every $u \in S$ there exists $\varepsilon > 0$ such that $B_\varepsilon(u) \subset S$. S is **closed** if its complement S^c is open.

Convergence in a Normed Space

Definition 2.17 Let $(\mathcal{U}, \|\cdot\|)$ be a normed space. We say that a sequence $[u_n]$ of elements of $\mathcal{U}$ converges to some $u \in \mathcal{U}$, if for every $\varepsilon > 0$ there exists a number M such that for every $n > M$ we have $\|u_n - u\| < \varepsilon$. In such a case we write

$$\lim_{n \to \infty} u_n = u$$

or simply $u_n \to u$.

This can be seen as convergence in real numbers: $u_n \to u$ if $\|u_n - u\| \to 0$ in $\mathbb{R}$

Definition 2.18 A function $f : \mathcal{U} \to \mathbb{R}$ is said to be continuous at $u \in \mathcal{U}$ if for any sequence $[u_m]$ with $u_m \to u$, we have $f(u_m) \to f(u)$ as $n \to \infty$. The function f is said to be continuous on $\mathcal{U}$ if it is continuous at every $u \in \mathcal{U}$

Proposition 2.1 *The norm function $f(u) = \|u\|$ is continuous*

Proof The idea is to show that if $u_m \to u$ then $\|u_m\| \to \|u\|$. The result follows by applying the property of the norm

$$|\|u\| - \|v\|| \leq \|u - v\| \; \forall u, v \in \mathcal{U}$$

Example The point evaluation functionals $\mathcal{U} \xrightarrow{\delta_{x_0}} \mathbb{R}$, $\delta_{x_0}(u) := u(x_0)$; $\mathcal{U} = C[0, 1]$ are continuous. Because, if we assume $u_m \to u$ in $\mathcal{U}$ as $m \to \infty$, then

$$|\delta_{x_0}(u_m) - \delta_{x_0}(u)| \leq \|u_m - u\|_\mathcal{U} \to 0 \text{ as } m \to \infty$$

Many norms can be defined in the same vector space. So it would be good for us to know if there is any relationship between them. We say two norms $\|\cdot\|_{(1)}$ and $\|\cdot\|_{(2)}$ and are equivalent if there exists constants c_1, c_2 such that

$$c_1 \|u\|_{(1)} \leq \|u\|_{(2)} \leq c_2 \|u\|_{(1)}, \; \forall u \in \mathcal{U}$$

With this property, if a sequence $[u_m]$ converge in one of the norms converge in the other norm

$$\|u_m - u\|_{(1)} \to 0 \iff \|u_m - u\|_{(2)} \to 0$$

If we consider the p-norms in $\mathbb{R}^N$

Under this point of view all the norms $\|\mathbf{x}\|_p$ $1 \leq p \leq \infty$ are equivalent on $\mathbb{R}^N$. It is not difficult to show that

$$\|\mathbf{x}\|_\infty \leq \|\mathbf{x}\|_p \leq N^{1/p} \|\mathbf{x}\|_\infty, \quad \forall \mathbf{x} \in \mathbb{R}^N$$

and as a consequence all the norms $\|\mathbf{x}\|_p$, $1 \leq p \leq \infty$ on $\mathbb{R}^N$ are equivalent. In fact we have the following result

Theorem 2.10 *Over a finite dimensional space, any two norms are equivalent*

Thus, in a finite-dimensional space, various norms result in the same concept of convergence. However, in an infinite-dimensional space, this statement does not hold true.

2.9 Analysis on Normed Spaces

Example If $\mathcal{U} = C[a, b]$, the space of continuous functions on [0, 1], the following norms can be defined on $\mathcal{U}$

$$\|u\|_p = \left[\int_a^b |u(x)|^p dx\right]^{1/p} \quad (2.26)$$

$$\|u\|_\infty = \sup_{a \le x \le b} |u(x)| \quad (2.27)$$

Considering the following sequence of functions $[u_m] \subset \mathcal{U}$ defined by

$$u_m(x) = \begin{cases} 1 - mx, & 0 \le x \le \frac{1}{m} \\ 0, & \frac{1}{m} \le x \le 1 \end{cases}$$

with some calculus

$$\|u_m\|_p = [m(p+1)]^{-1/p}, 1 \le p < \infty.$$

and

$$\|u_m\|_\infty = 1, \qquad m \ge 1$$

Then $[u_m] \to 0$ in the norm $\|\cdot\|_p$, $1 \le p < \infty$ but does not converge to 0 in the norm $\|\cdot\|_\infty$. So in infinite dimensional spaces some norms are not equivalent. However one norm may be stronger than other, for example, it can be shown that

$$\|u_m\|_p \le \|u_m\|_\infty, \forall u \in \mathcal{U}$$

Therefore convergence in $\|\cdot\|_\infty$ implies convergence in $\|\cdot\|_p$ but not conversely. Usually, convergence in $\|\cdot\|_\infty$ is called *uniform convergence* □

In the original definition of convergence it is assumed that the limit of the sequence is known. However, in many cases this is not possible. In the search for a more operational notion of the concept of convergence, we arrive at the definition of Cauchy sequence.

Definition 2.19 Let $\mathcal{U}$ be a normed space. A sequence $[u_k] \subset \mathcal{U}$ is called a Cauchy sequence if

$$\lim_{m,n \to \infty} \|u_m - u_n\| = 0$$

It is easy to show that a convergent sequence is a Cauchy sequence, thus it is a necessary condition for convergence. The following step is to ask if any Cauchy sequence is convergent; this is certain in $\mathbb{R}^N$ but not in general for infinite dimensional spaces.

Banach Spaces

Definition 2.20 A normed linear space $\mathcal{U}$ is said to be *complete* if every Cauchy sequence in $\mathcal{U}$ converges to an element in $\mathcal{U}$. A complete normed space is called a **Banach space**.

Example The space $C[a, b]$ with the maximum norm is a Banach space. To show this let u_n be a Cauchy sequence in $C[a, b]$. Then for sufficiently large n and p $\|u_{n+p}(t) - u_n(t)\| < \varepsilon$ for all $t \in [a, b]$. From a standard result in advanced calculus [101] it then follows that the sequence converges uniformly to a limit which is continuous. □

Example The space $C[a, b]$ with the norm

$$\|u\|_2 := \left(\int_a^b u^2(t)dt\right)^{1/2}$$

is not complete. For example in $[-1, 1]$ the sequence (Fig. 2.7)

$$u_n(t) = \begin{cases} -1, & -1 \le t \le -\frac{1}{n} \\ nt, & -\frac{1}{n} < t < \frac{1}{n} \\ 1, & \frac{1}{n} \le t \le 1 \end{cases} \qquad (2.28)$$

is a Cauchy sequence but its pointwise limit is not continuous □

If we define the space $\mathcal{L}^p(\Omega)$ as the set of measurable functions in the sense of Lebesgue such that

$$\|u\|_p = \left(\int_\Omega |u(x)|^p dx\right)^{1/p} < \infty, \qquad (2.29)$$

To construct a complete space, one must include all Lebesgue integrable functions, which leads to $\mathcal{L}^2$ being a Banach space with the given norm. This result is well-established in analysis. The spaces $\mathcal{L}^p(\Omega)$, $1 \le p \le \infty$, are Banach spaces, and concrete realizations of the abstract completion of $C(\bar{\Omega})$ under the norm (2.29).

Fig. 2.7 A graph of the sequence (2.28)

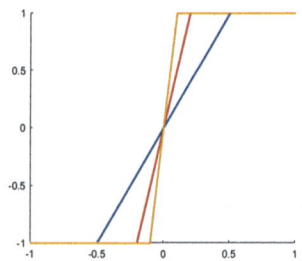

The space $\mathcal{L}^\infty(\Omega)$ is a Banach space as well, but much larger than $C(\bar{\Omega})$ with $\|\cdot\|_\infty$. The requirement for interpreting integrals as Lebesgue integrals arises from this completeness, as the collection of Riemann-integrable functions is not complete with respect to the p-norm.

2.10 Inner Product and Hilbert Spaces

Normed linear spaces whose norm is induced by an inner product have several additional important properties which we now describe in detail.

Let $\mathcal{U}$ be a linear space over $\mathbb{K} = \mathbb{R}$ or $\mathbb{C}$, then the inner product $\langle u, v \rangle$ of $u, v \in \mathcal{U}$ is an operation $\langle \cdot, \cdot \rangle : \mathcal{U} \times \mathcal{U} \to \mathbb{K}$ that satisfies the following properties for all $u, v, w \in \mathcal{U}$ and $\alpha, \beta \in \mathbb{R}$

1. $\langle u, v \rangle = \overline{\langle v, u \rangle}$ (Symmetry)
2. $\langle u, v + w \rangle = \langle u, v \rangle + \langle u, w \rangle$ (Additivity)
3. $\langle \lambda u, v \rangle = \lambda \langle u, v \rangle$ (Homogeneity)
4. $\langle u, u \rangle \geq 0$ and $\langle u, u \rangle = 0$ if and only if $u = 0$ (Positive Definitness)

The space $\mathcal{U}$ together with the inner product $\langle \cdot, \cdot \rangle$ is called an inner product space. For the case of real inner product the second axiom reduces to $\langle u, v \rangle = \langle v, u \rangle$ $\forall u, v \in \mathcal{U}$.

If $\mathcal{U}$ is an inner product space it holds the **Schwartz inequality**

$$|\langle u, v \rangle| \leq \sqrt{\langle u, u \rangle \langle v, v \rangle} \qquad \forall u, v \in \mathcal{U}$$

and the inner product induces a norm by just defining

$$\|u\| := \sqrt{\langle u, u \rangle} \qquad \forall u \in \mathcal{U}$$

Example The canonical inner product in $\mathbb{R}^N$ was already defined in (2.3). In the same way an inner product can be defined in $\mathbb{C}^N$; for $\mathbf{x} = [x_1, \cdots, x_N]^T$ and $\mathbf{y} = [y_1, \cdots, y_N]^T$

$$\langle \mathbf{x}, \mathbf{y} \rangle = \sum_{k=1}^N x_k \overline{y_k} = \mathbf{y}^* \mathbf{x}$$

and

$$\|\mathbf{x}\| = \sqrt{\langle \mathbf{x}, \mathbf{x} \rangle} = \left(\sum_{k=1}^N |x_k|^2 \right)^{1/2}$$

Example The linear space $\mathcal{U} = \mathbb{R}^N$ can be given an inner product structure by

$$\langle \mathbf{x}, \mathbf{y} \rangle_A := \langle A\mathbf{x}, \mathbf{y} \rangle = \sum\sum a_{ij} x_i y_j$$

If A is a positive definite matrix

Example The space $\mathcal{L}^2(\Omega)$ is an inner product space with the canonical inner product

$$\langle u, v \rangle = \int_\Omega u(x)\overline{v(x)}$$

and the induced standard $\mathcal{L}^2$-norm

$$\|u\| = \sqrt{\langle u, u \rangle} = \left(\int_\Omega |u(x)|^2 dx \right)^{1/2}$$

Definition 2.21 A complete inner product space is called a **Hilbert space**

Equivalently, an inner product space $\mathcal{U}$ is a Hilbert space if $\mathcal{U}$ is a Banach space under the induced norm.

Example

- $\mathbb{C}^N$ is a Hilbert space with the inner product

$$\langle \mathbf{x}, \mathbf{y} \rangle = \sum_{k=1}^{N} x_k \overline{y_k}$$

- as well as the space l^2 formed by the sequences $\mathbf{x} = [x_k]_{k \geq 1}$ such that $\sum_{k}^{\infty} |x_k|^2 < \infty$. In this space

$$\alpha \mathbf{x} + \beta \mathbf{y} = [\alpha x_k + \beta y_k]_{k \geq 1}$$

and the inner product

$$\langle \mathbf{x}, \mathbf{y} \rangle = \sum_{k=1}^{\infty} x_k \overline{y_k}$$

- The space $\mathcal{L}^2(\Omega)$ is a Hilbert space under the inner product

$$\langle u, v \rangle = \int_\Omega u(x)\overline{v(x)}$$

2.10 Inner Product and Hilbert Spaces

Definition 2.22 (Orthogonality) Let $(\mathcal{U}, \langle, \rangle)$ be an inner product space
- Two vectors are said to be orthogonal ($u \perp v$) if $\langle u, v \rangle = 0$.
- The vector $v \in \mathcal{U}$ is said to be orthogonal to $U \subset \mathcal{U}$ if $v \perp u$ for any $u \in U$
- A set $U \subset \mathcal{U}$ is called an *orthogonal system* if its elements are mutually orthogonal. If $u_1, \cdots, u_M$ are mutually orthogonal

$$\left\| \sum_k u_k \right\|^2 = \sum_k \|u_k\|^2$$

- For a subspace $U \in \mathcal{U}$, the orthogonal complement of U is the subspace

$$U^\perp = \{v \in \mathcal{U} : v \perp u \quad \forall u \in \mathcal{U}\}$$

Theorem 2.11 *Let $\mathcal{U}$ be Hilbert Space and u_k an orthonormal sequence in $\mathcal{U}$. Then the following are equivalent*

(i) u_k is an orthonormal sequence
(ii) for every $X \in \mathcal{U}$

$$X = \sum_{k=1}^{\infty} (X, u_k) u_k$$

(iii) for every $w \in \mathcal{U}$,

$$\|w\|^2 = \sum_{k=1}^{\infty} |\langle w, u_k \rangle|^2$$

Definition 2.23 (Adjoint Operator) Given a linear operator $\mathscr{A} : \mathcal{U} \to \mathcal{V}$ between two inner product spaces $\mathcal{U}, \mathcal{V}$, the **adjoint** of $\mathscr{A}$, denoted $\mathscr{A}^*$, is a linear operator $\mathscr{A}^* : \mathcal{V} \to \mathcal{U}$ such that for all $u \in \mathcal{U}, v \in \mathcal{V}$, the following condition holds:

$$\langle \mathscr{A}u, v \rangle_\mathcal{V} = \langle u, \mathscr{A}^* v \rangle_\mathcal{U}$$

If $\mathscr{A} = \mathscr{A}^*$, $\mathscr{A}$ is called to be self-adjoint.

Example If $\mathcal{U} = \mathbb{R}^N$, linear maps $\mathscr{A} : \mathbb{R}^N \to \mathbb{R}^N$ are square matrices, then

$$\langle \mathscr{A}\mathbf{x}, \mathbf{y} \rangle = (\mathscr{A}\mathbf{x})^T \mathbf{y} = \mathbf{x}^T (\mathscr{A}^T \mathbf{y}) = \langle \mathbf{x}, \mathscr{A}^T \mathbf{y} \rangle$$

as a consequence $\mathscr{A}^* = \mathscr{A}^T$. □

Example Observe the formula (2.22) of integration by parts can be written applying the inner product as

$$\left\langle \frac{\partial u}{\partial x_i}, v \right\rangle = \left\langle u, -\frac{\partial v}{\partial x_i} \right\rangle + \int_{\partial \Omega} u v v_i \, dS$$

If v vanishes on $\partial \Omega$ then the integral on $\partial \Omega$ vanishes and

$$\left\langle \frac{\partial u}{\partial x_i}, v \right\rangle = \left\langle u, -\frac{\partial v}{\partial x_i} \right\rangle$$

This is saying that the adjoint of the linear operator $\frac{\partial}{\partial x_i}$ is $-\frac{\partial}{\partial x_i}$. □

Chapter 3
Linear Operators and Functionals

In this chapter, we explore the foundational role of linear operators and functionals in the development of modern mathematical frameworks such as distributions or generalized functions [62]. Starting with an introduction to bounded linear functionals and the Riesz representation theorem, we establish the core principles that lead to the formalization of distributions. Using the intuitive concept of integration by parts, we extend the definition of the derivative to include discontinuous functions, with particular attention to the Dirac delta function. We then formally define distributions as linear functionals acting on a space of test functions and demonstrate how they can be used to represent images in image processing. The final section of the chapter focuses on Fourier analysis, applying the theory to problems such as convolution and Fourier transforms, offering practical tools for solving real-world imaging problems [42, 70, 100, 107].

3.1 Linear Operators

Let $\mathcal{U}, \mathcal{V}$ be linear spaces, an operator $\mathcal{A}$ from $\mathcal{U}$ to $\mathcal{V}$ is a rule which assigns to each element in a subset of $\mathcal{U}$ a unique element in $\mathcal{V}$. The domain $D(\mathcal{A})$ of $\mathcal{A}$ is the subset of $\mathcal{U}$ where $\mathcal{A}$ is defined. Usually the domain is assumed to be the whole set $\mathcal{U}$. The operator $\mathcal{A} : \mathcal{U} \to \mathcal{V}$ is linear if

(a) for all $u, v \in \mathcal{U}$,

$$\mathcal{A}(u+v) = \mathcal{A}(u) + \mathcal{A}(v)$$

(b) for all $u \in \mathcal{U}$ and $\alpha \in \mathbb{R}$

$$\mathcal{A}(\alpha u) = \alpha \mathcal{A}(u)$$

For a linear operator $\mathscr{A}$ is usually written $\mathscr{A}u$ for $\mathscr{A}(u)$.

It is easy to show that the set of all linear operators defined between two vector spaces $\mathcal{U}$ and $\mathcal{V}$ form a vector space themselves. This vector space is denoted by $L(\mathcal{U}, \mathcal{V})$, where the operations of addition and scalar multiplication are defined in the obvious way.

Bounded Linear Operators If $\mathcal{U}, \mathcal{V}$ are normed spaces, the linear operator $L : \mathcal{U} \to \mathcal{V}$ is called *bounded* if there exists a number $\alpha \in \mathbb{R}$ such that for all elements $u \in \mathcal{U}$

$$\|L(u)\| \leq \alpha \|u\|.$$

For each bounded, linear operator we define a norm as

$$\|L\| := \sup_{u \neq 0} \frac{\|Lu\|}{\|u\|}; \quad \forall u \in D(L) \tag{3.1}$$

The set of all bounded linear operators from $\mathcal{U}$ into $\mathcal{V}$ is denoted by $\mathcal{L}(\mathcal{U}, \mathcal{V})$. Every linear combination of bounded linear operators again is a bounded linear operator, i.e., the set $\mathcal{L}(\mathcal{U}, \mathcal{V})$ is a linear space.

- If a linear operator is continuous in element $u_0 \in \mathcal{U}$ then the operator is continuous for all $u \in \mathcal{U}$. This can be seen in the following way. Let $\mathscr{A}$ be continuous at $u_0 \in \mathcal{U}$. Then for every $u \in \mathcal{U}$ and every sequence (u_k) with $u_k \to u, k \to \infty$, we have

$$\mathscr{A}u_k = A(u_k - u + u_0) + A(u - u_0) \to \mathscr{A}(u_0) + \mathscr{A}(u - u_0) = \mathscr{A}(u), k \to \infty$$

since $(u_k - u + u_0) \to u_0, k \to \infty$. Therefore, $\mathscr{A}$ is continuous at all $u \in \mathcal{U}$.
- Let $\mathcal{U}, \mathcal{V}$ be normed spaces and $\mathscr{A} : \mathcal{U} \to \mathcal{V}$ a linear operator. If $\mathcal{U}$ is finite dimensional then $\mathscr{A}$ is bounded
- Let $\mathcal{U}, \mathcal{V}, \mathcal{W}$ be normed spaces and $\mathscr{A} : \mathcal{U} \to \mathcal{V}$ and $\mathscr{B} : \mathcal{V} \to \mathcal{W}$ bounded linear operators. Then the product $\mathscr{B}\mathscr{A} : \mathcal{U} \to \mathcal{W}$ defined by $(\mathscr{B}\mathscr{A})u := \mathscr{B}(\mathscr{A}u)$ for all $u \in \mathcal{U}$ is a bounded linear operator with $\|\mathscr{B}\mathscr{A}\| \leq \|\mathscr{A}\| \|\mathscr{B}\|$

Example (Integral Operators) An important case of bounded linear operators are defined in the following way. Let $\Omega \subset \mathbb{R}^N$ be a nonempty compact set and $K : \Omega \times \Omega \to \mathbb{R}$ a continuous function. Then the linear operator $\mathscr{A} : \Omega \to \Omega$ defined by

$$(\mathscr{A}u)(x) := \int_\Omega K(x, y) u(y) dy, \qquad x \in \Omega$$

is called an integral operator with continuous kernel K. $\mathscr{A}$ is a bounded linear operator with

$$\|\mathscr{A}\|_\infty = \max_{x \in \Omega} \int_\Omega |K(x, y)| dy$$

3.2 Linear Functionals

Perhaps the most frequent use of the normed vector spaces $L(\mathcal{U}, \mathcal{V})$ arises when $\mathcal{V}$ is chosen as $\mathbb{R}$, endowed with the usual topology. The space $L(\mathcal{U}, \mathbb{R})$ of all linear mappings from $\mathcal{U}$ into $\mathbb{R}$, is called the algebraic dual $\mathcal{U}^\dagger$ of $\mathcal{U}$. The elements of $\mathcal{U}^\dagger$ are called linear functionals.

Definition 3.1 A real valued mapping defined on a normed linear space is called a **functional**. If the mapping is linear, it is called a **linear functional**.

The space $\mathcal{U}^\dagger = L(\mathcal{U}, \mathbb{R})$ is called the algebraic dual of $\mathcal{U}$ and the set of bounded linear functionals endowed with the operator norm is the **topological dual space** $\mathcal{U}^* := \mathcal{L}(\mathcal{U}, \mathbb{R})$ of $\mathcal{U}$.

Example If u is a function in $\mathcal{U}$ then $u : U \to \mathbb{R}$ for some set U. Then we can choose an element of U, say x_0, and evaluate u at x_0 to obtain $u(x_0)$, where u is any element of $\mathcal{U}$. We define the **point evaluation functionals** as

$$\delta_{x_0} : \mathcal{U} \to \mathbb{R}; \qquad \delta_{x_0}(u) := u(x_0)$$

If $\mathcal{U}$ is a linear space of functions with pointwise addition and scalar multiplication we can see

$$\delta_{x_0}(\alpha u + \beta v) = (\alpha u + \beta v)(x_0) = \alpha u(x_0) + \beta v(x_0) = \delta_{x_0}(u) + \delta_{x_0}(v).$$

then δ_{x_0} is a linear functional.

□

Example Let $w_1, w_2, \cdots, w_N$ be any real numbers and let $\mathcal{U}$ be a linear space of functions; then the finite sum

$$S(u) = \sum_{k=1}^{N} w_k u(x_k)$$

is a linear functional on $\mathcal{U}$ for any choice of the arguments $x_1, x_2, \cdots, x_N$ We can view this linear functional as a linear combination of evaluation functionals.

□

Example If $\mathcal{U}$ is the linear space of real valued integrable functions on $[a, b]$, then the definite integral

$$J(u) = \int_a^b u(t)dt$$

is a linear functional on $\mathcal{U}$. We have $J(\alpha u + \beta v) = \alpha J(u) + \beta J(v)$.

□

For $\mathcal{U} = C[a,b]$, the three examples above are bounded linear functionals. In fact,

- $\delta_t(u) \leq \|u\|$
- $|S(u)| \leq \left(\sum |w_k|\right) \|u\|$
- $|J(u)| \leq (b-a)\|u\|$

Example If $\mathcal{U}$ is an inner product space and h is any element of $\mathcal{U}$, then

$$P_h(u) = \langle u, h \rangle$$

is a linear functional on $\mathcal{U}$.

Example In the linear space $\mathcal{U} = C[0,1]$ consider the functional

$$F(u) := \int_0^1 fu\, dt$$

where f is integrable. Notice that

$$F(\alpha u + \beta v) = \alpha F(u) + \beta F(v)$$

and

$$|F(u)| \leq \max_t |u(t)| \int_0^1 |f(t)|dt$$

so F is a bounded linear functional.

$\square$

Example (Linear Functionals in $\mathbb{R}^N$) Any linear functional in $\mathbb{R}^N$ has the form $\boldsymbol{a} \cdot \mathbf{x}$. This can be seen in the following way. If $\alpha_1, \cdots, \alpha_n$ are real numbers and $\mathbf{x} \in \mathbb{R}^N$ the function defined by

$$L(\mathbf{x}) = \alpha_1 x_1 + \cdots, \alpha_n x_N = \boldsymbol{a}^T \mathbf{x} \qquad (3.2)$$

is linear. On the other side, if $\mathbf{x} \in \mathbb{R}^N$, L is linear and $\mathbf{x} = \sum \alpha_j e_j$ then

$$L(\mathbf{x}) = L\left(\sum \alpha_j e_j\right) = \sum \alpha_j L(e_j) = \boldsymbol{\alpha}^T w$$

$w = [L(e_1), \cdots, L(e_N)]^T$. This means that if L is a linear functional in $\mathbb{R}^N$ there exists an element $\boldsymbol{a}$ in $\mathbb{R}^N$ such that $L(\mathbf{x}) = \boldsymbol{a} \cdot \mathbf{x} = \boldsymbol{a}^T \mathbf{x}$.

$\square$

3.3 Continuous Linear Functionals

A continuous linear functional $\mathscr{L}$ on a normed space $\mathcal{U}$ is a linear map

$$\mathscr{L} : \mathcal{U} \to \mathbb{R}(\text{or } \mathbb{C})$$

that is continuous with respect to the norm topology on $\mathcal{U}$. For a linear functional $\mathscr{L} : \mathcal{U} \to \mathbb{R}$ the following statements are equivalent.

1. $\mathscr{L}$ is continuous at one $u_0 \in \mathcal{U}$
2. $\mathscr{L}$ is continuous on $\mathcal{U}$
3. $\mathscr{L}$ is bounded on $\mathcal{U}$

It is common to regard the application of elements in $\mathcal{U}^*$ to elements in $\mathcal{U}$ as a bilinear mapping called duality pairing:

$$\langle \cdot, \cdot \rangle_{\mathcal{U}^* \times \mathcal{U}} : \mathcal{U}^* \times \mathcal{U} \to \mathbb{R}$$

and frequently is written as

$$u^*(u) := \langle u^*, u \rangle_{\mathcal{U}^* \times \mathcal{U}} \tag{3.3}$$

Continuous linear functionals form the backbone of distribution theory, allowing the extension of classical function spaces to include generalized functions. This extension is crucial for many applications in mathematics and physics, particularly in dealing with singularities and providing a rigorous foundation for solving PDEs and other complex problems.

Riesz Representation Theorem

The Riesz Representation Theorem is a cornerstone result in functional analysis, particularly in the study of Hilbert spaces. It provides a deep connection between a Hilbert space and its dual space by showing that every continuous linear functional on a Hilbert space can be represented as an inner product with a fixed element from the space.

Statement of the Theorem Let $\mathcal{H}$ be a Hilbert space, and let $\mathscr{L} : \mathcal{H} \to \mathbb{R}$ be a continuous linear functional on $\mathcal{H}$. Then there exists a unique element $z \in \mathcal{H}$ such that

$$\langle \mathscr{L}, u \rangle = \mathscr{L}(u) = \langle u, z \rangle \qquad \forall u \in \mathcal{H} \tag{3.4}$$

The element z is sometimes called the **representer** of the functional $\mathscr{L}$.

Example If $\mathcal{H} = \mathbb{R}^N$ any linear functional $\mathscr{L}: \mathbb{R}^N \to \mathbb{R}$ has the form

$$\mathscr{L}(\mathbf{x}) = \sum_{i=1}^{n} w_i x_i = \langle \mathbf{w}, \mathbf{x} \rangle$$

Here, the Riesz representation theorem tells us that for every linear functional, there is a corresponding vector $\mathbf{w}$ such that the functional is the dot product with $\mathbf{w}$ □

Example The dual space of $\mathcal{L}^p$, denoted by $(\mathcal{L}^p)^*$, can be identified with $\mathcal{L}^{p'}$ where $1/p + 1/p' = 1$ and $1 \leq p < \infty$ [101]. Other form to see the theorem is as follows: If $\mathscr{L}$ is a continuous linear functional on $\mathcal{L}^p$, then there exists a unique $u \in \mathcal{L}^{p'}$ such that

$$\langle \mathscr{L}, f \rangle = \mathscr{L}(f) = \int_\Omega uf, \qquad \forall f \in \mathcal{L}^p$$

and moreover

$$\|u\|_{\mathcal{L}^{p'}} = \|\mathscr{L}\|_{(\mathcal{L}^p)^*}$$

3.4 Distributions

One way to enter the world of distributions is by following the history of Dirac's δ and the mechanisms designed to overcome or explain the contradictions in the definition given by Dirac [31, 32, 120]:

- $\delta(x) = \begin{cases} \infty; x = 0 \\ 0, \text{otherwise} \end{cases}$
- $\int \delta(x) = 1$
- $\int \delta(x)\varphi(x) = \varphi(0)$

Dirac's δ also arises when we want to calculate the derivative of functions that contain jumps, such as the Heaviside function $H(x)$

$$H(x) = \begin{cases} 1, & x > 0 \\ 0, & x < 0 \end{cases} \qquad (3.5)$$

In this type of situation we can observe that functions with jumps cannot be differentiated but they can be integrated. Therefore, a possible solution would be to use the properties of integration appropriately to generalize the derivative of a function. In fact that is the way of Distributions theory, by using the properties of integration by parts. If in the formula

3.4 Distributions

$$\int_a^b f\varphi' = [f\varphi]_a^b - \int_a^b f'\varphi \tag{3.6}$$

we consider functions φ such that $\varphi(a) = \varphi(b) = 0$, then

$$\int_a^b f\varphi' = -\int_a^b f'\varphi, \tag{3.7}$$

and if we want to avoid problems with the integration interval we consider integrals in $[-\infty, \infty]$

$$\int_{-\infty}^{\infty} f'\varphi = -\int_{-\infty}^{\infty} f\varphi'$$

or in duality notation

$$\langle f', \varphi \rangle = -\langle f, \varphi' \rangle \qquad \text{Derivative generalization} \tag{3.8}$$

Since f' appears on the left side of the equation, this formula could be taken as a definition of the derivative of a function [1]. In other words, we have transferred the problem of derivatives to the world of certain linear functionals whose domain is made up of a very particular set $C_0^\infty(\mathbb{R}^N)$ of functions φ, called **test functions**. These functions have a certain value on a finite interval, vanish outside that interval and are infinitely differentiable. Therefore they have no choice but to have the shape of a "bump" (Fig. 3.1). A classical example is the function

$$\varphi_a(\mathbf{x}) = \begin{cases} e^{-\frac{a^2}{a^2 - \|\mathbf{x}\|^2}}, & \|\mathbf{x}\| < a \\ 0, & \text{otherwise} \end{cases} \tag{3.9}$$

The symbol $\langle f, \varphi \rangle$ does not assign pointwise values of the function f, but rather describes the action of f on the bump functions. Under this viewpoint the problem with Dirac's δ is solved, defining its action on the set of test functions $C_0^\infty(\mathbb{R})$ (Fig. 3.2). Then δ is defined as

$$\langle \delta, \varphi \rangle := \varphi(0) \tag{3.10}$$

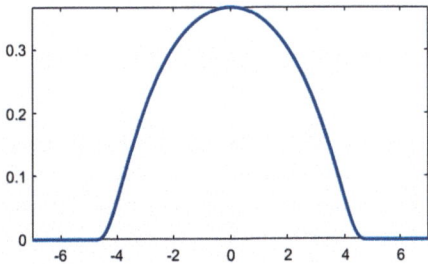

Fig. 3.1 Typical test function (3.9) $\varphi_a(\mathbf{x})$ for $a = 5$ and $\mathbf{x} \in \mathbb{R}$

Fig. 3.2 Test function $\varphi_a(\mathbf{x})$ for $a = 5$ and $\mathbf{x} \in \mathbb{R}^2$

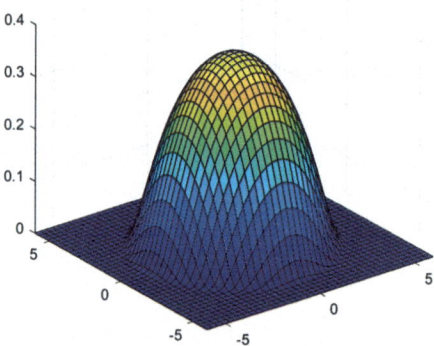

So if we need to find the derivative of the Heaviside function (3.5) we can say

$$\langle H', \varphi \rangle = -\langle H, \varphi' \rangle = -\int_{-\infty}^{+\infty} H(x)\varphi' = -\int_{-\infty}^{0} 0 - \int_{0}^{+\infty} 1 \cdot \varphi'$$

$$= -\int_{0}^{+\infty} \varphi dx = \varphi(0) - \overbrace{\varphi(\infty)}^{0} = \varphi(0) = \langle \delta, \varphi \rangle$$

and finally it makes sense to write $H' = \delta$. The same reasoning can be applied to define $\delta_{(a)}$ (or $\delta(x - a)$) as

$$\langle \delta_{(a)}, \varphi \rangle := \varphi(a)$$

It should be remarked that although it is not formally correct, the notation $\int \delta(t - a) f \, dt$ is commonly used for referring to $\langle \delta_a, f \rangle$.

From here on an integral sign with no limits will denote integration over all of N-space:

$$\int f(\mathbf{x}) \, d\mathbf{x} \quad \text{means} \quad \int_{-\infty}^{\infty} \cdots \int_{-\infty}^{\infty} f(x_1, ..., x_N) dx_1 \cdots dx_N$$

3.5 Test Functions

Multi-Index Notation

The class $C^k(\Omega)$ consists of the real valued functions on Ω which have continuous derivatives of order less than, or equal to k (f is included as derivative of order zero). In the same way $C^\infty(\Omega)$ consists of the functions which have continuous derivatives of all orders.

3.5 Test Functions

The support of a function $f : \Omega \to \mathbb{R}$ is the closure of the set $\{x \in \Omega : f(x) \neq 0\}$ note that it is a closed subset of Ω. We shall write $\operatorname{supp} f = \overline{\{x \in \Omega : f(x) \neq 0\}}$. Functions with compact support play an important part in the theory. We write C_0^k for the subset of C^k consisting of functions with compact support.

An element $\alpha \in \mathbb{Z}_n^+$, called **multi-index**, is an ordered collection of n non-negative integers

$$\alpha = (\alpha_1, \cdots, \alpha_n)$$

We denote by $|\alpha|$ the sum $|\alpha| = \sum_{i=1}^n \alpha_i$ and by $D^\alpha u$ the partial derivative

$$D^\alpha u = \frac{\partial^{|\alpha|} u}{\partial x_1^{\alpha_1} \partial x_2^{\alpha_2} \ldots \partial x_n^{\alpha_n}}$$

Example

- If $\mathbf{x} \in \mathbb{R}^N$ and $\alpha \in \mathbb{Z}_3^+$ with $\alpha = (3, 0, 2)$ then $|\alpha| = \alpha_1 + \alpha_2 + \alpha_3 = 3 + 0 + 2 = 5$. The derivative $D^\alpha u$ is

$$D^\alpha u = \frac{\partial^5 u}{\partial x_1^{\alpha_1} \partial x_2^{\alpha_2} \partial x_n^{\alpha_3}} = \frac{\partial^5 u}{\partial x^3 \partial y^0 \partial z^2} = \frac{\partial^5 u}{\partial x^3 \partial z^2}$$

- If $n = 2$ Consider the expression

$$\mathcal{P} = \sum_{|\alpha| \leq 2} c_\alpha D^\alpha u$$

with functions $c_\alpha(x, y)$. Then

$$\mathcal{P} = \sum_{|\alpha|=0} c_\alpha D^\alpha u + \sum_{|\alpha|=1} c_\alpha D^\alpha u + \sum_{|\alpha|=2} c_\alpha D^\alpha u$$

Definition 3.2 ([35, 60, 130]) The space of test function, denoted $\mathcal{D}$ or C_0^∞ is the space of all real-valued functions $\varphi(\mathbf{x})$, $\mathbf{x} \in \mathbb{R}^N$ such that the following hold:

1. $\varphi(\mathbf{x})$ is in $C^\infty(\mathbb{R}^N)$ is an infinitely differentiable function defined on $\mathbb{R}^N$. This means that $D^\alpha \varphi$ exists for all multiindices α.
2. There exists a number A such that $\varphi(\mathbf{x})$ vanishes for $r > A$. This means that $\varphi(\mathbf{x})$ has compact support. Then $\varphi(\mathbf{x})$ is called a test function.

Test functions have the following properties, easy to verify

1. If $\varphi_1, \varphi_2 \in \mathcal{D}$ then so is $c_1 \varphi_1 + c_2 \varphi_2$, $c_1, c_2 \in \mathbb{R}$. Thus $\mathcal{D}$ is a linear space.
2. If $\varphi \in \mathcal{D}$ then $D^\alpha \varphi \in \mathcal{D}$
3. For $f \in C^\infty$ and for $\varphi \in \mathcal{D}$, $f\varphi \in \mathcal{D}$

A function $f(\mathbf{x})$ is locally integrable in $\mathbb{R}^N$ if $\int_\Omega |f(\mathbf{x})|\,d\mathbf{x}$ exists for every bounded region Ω in $\mathbb{R}^N$. By $\mathcal{L}^1_{loc}(\Omega)$ we denote the space of functions whose absolute value is integrable over every compact subset of Ω.

The Space $C_0^\infty(\Omega)$ and Integration by Parts

The formulas for integration by parts can be stated in terms of the space $C_0^\infty(\Omega)$

(i) The integration by parts formula (3.6) holds for all

$$f, \varphi \in C^1[a,b]$$

(ii) Formula (3.7) holds for all

$$f \in C^1[a,b] \quad \text{and} \quad \varphi \in C_0^\infty(a,b)$$

(iii) The generalization of the integration by parts formula (3.6) to $\mathbb{R}^N$ is:

$$\int_\Omega \frac{\partial f}{\partial x_j} \varphi d\mathbf{x} = \int_{\partial\Omega} f\varphi n_j dS - \int_\Omega f \frac{\partial \varphi}{\partial x_j} \qquad \text{for all } f, \varphi \in C^1(\bar{\Omega}) \tag{3.11}$$

provided Ω is a nonempty bounded open set in $\mathbb{R}^N$ that has sufficiently smooth boundary. In the two-dimensional case ($N = 2$) the surface integral $\int_\Omega \cdots dS$ is taken in the sense of $\int_{\partial\Omega} \cdots ds$, with s arclength, and the boundary curve $\partial\Omega$ oriented in such a manner that the set Ω lies on the left-hand side of $\partial\Omega$. The generalization of (3.7) is

(iv)

$$\int_\Omega \frac{\partial f}{\partial x_j} \varphi\, d\mathbf{x} = -\int_\Omega f \frac{\partial \varphi}{\partial x_j}\, d\mathbf{x} \qquad \text{for all } f \in C^1(\bar{\Omega});\ \varphi \in C_0^\infty(\bar{\Omega}) \tag{3.12}$$

provided Ω is a nonempty open set in $\mathbb{R}^N$.

The Space $C_0^\infty(\Omega)$ and the Variational Lemma

The following lemma plays an important role in the Calculus of Variations

Lemma 3.1 (Fundamental Lemma of Variational Calculus ([129])) *Let Ω a nonempty open set in $\mathbb{R}^N$. Then, it follows from $u \in \mathcal{L}^2(\Omega)$ and*

3.5 Test Functions

$$\int_\Omega u\varphi\, d\mathbf{x} = 0 \quad \text{for all } \varphi \in C_0^\infty(\Omega) \tag{3.13}$$

that $u(\mathbf{x}) = 0$ for almost all $\mathbf{x} \in \Omega$.

If, in addition, $u \in C(\Omega)$, then $u(\mathbf{x}) = 0$ for all $\mathbf{x} \in \Omega$.

The properties of integration by parts can be used to generalize the notion of a derivative. For example, the definition of the weak derivative, which generalizes the classical derivative to functions that may not be differentiable in the classical sense but are still integrable.

Definition 3.3 If $u \in \mathcal{L}_{loc}^1(\Omega)$ is a locally integrable function, a function $v \in \mathcal{L}_{loc}^1(\Omega)$ is called the **weak derivative** of u if

$$\int_\Omega u \frac{\partial \varphi}{\partial x_i} d\mathbf{x} = -\int_\Omega v\varphi d\mathbf{x} \qquad \forall \varphi \in C_0^\infty \tag{3.14}$$

Example Consider the function $u(x) = |x|$ on $\Omega =]0, 1[$. Its classical derivative is

$$u'(x) = \begin{cases} 1, & x > 0, \\ -1, & x < 0, \\ \text{undefined}, & x = 0. \end{cases}$$

The weak derivative v must satisfy:

$$\int_{-1}^{1} |x|\varphi'(x) = -\int_{-1}^{1} v\varphi$$

By integrating by parts and recognizing the behavior at $x = 0$ we find:

$$v(x) = \begin{cases} 1, & x > 0, \\ -1, & x < 0 \end{cases}$$

This v is the weak derivative of $u(x) = |x|$

$\square$

Definition 3.4 A functional $T : C_0^\infty(\Omega) \to \mathbb{R}$ is said to be a **distribution** (or **generalized function**) (on Ω) if J is linear and continuous, i.e.,

- $T(\alpha_1\varphi_1 + \alpha_2\varphi_2) = \alpha_1 T(\varphi_1) + \alpha_2 T(\varphi_2)$
- $\varphi_k \to \varphi$ in $C_0^\infty(\Omega)$ implies $T(\varphi_k) \to T(\varphi)$ in $\mathbb{R}$

This is, distributions are continuous linear functionals that act on the space of test functions. The set of all distributions on Ω is symbolized $\mathcal{D}'(\Omega)$. Obviously, $\mathcal{D}'(\Omega)$ is a vector space [45].

The space $\mathcal{L}_{loc}^1(\mathbb{R}^N)$ generate a very useful set of distributions. Indeed, if $f(\mathbf{x})$ is locally integrable, defines a distribution through the formula

$$\langle f, \varphi \rangle := \int_{\mathbb{R}^N} f(\mathbf{x}) \varphi(\mathbf{x}) \, d\mathbf{x} \tag{3.15}$$

witch is linear functional. To prove that this functional is continuous

$$|\langle f, \varphi \rangle| \leq \max_{\mathbf{x} \in \mathrm{supp}\varphi} |\varphi(\mathbf{x})| \int_{\mathrm{supp}\varphi} |f(\mathbf{x})| \, d\mathbf{x}$$

If the sequence $\{\varphi_m\} \to 0$ then $\langle f, \varphi \rangle \to 0$ (see [108]). Hence, it is continuous. Distributions defined by (3.15) are called **regular**. All other distributions are called **singular**.

Representation of Images by Distributions

Note that the function representation treats images as continuous entities, ideal for tasks that rely on smoothness and differential calculus, but a digital image can be represented as a distribution, particularly when it involves sharp edges, noise, or abrupt changes that make a traditional smooth function representation inadequate.

The idea that a test function can be interpreted as a linear sensor for capturing image signals provides an interesting bridge between distribution theory and image processing. In this framework, we treat an image as a signal that is "sampled" or "sensed" by a test function, resulting in a measurement or observation. If $u(\mathbf{x})$ represents the intensity of an image at a point $\mathbf{x}$, then $\langle u, \varphi \rangle = \int u\varphi$ represents the response of the image u to the test function φ.

Convolution $(u * \varphi)(x) = \int u(t)\varphi(x - t)dt$ of an image u with a filter φ can be interpreted as applying a test function (or kernel) to capture features at different scales or positions. Here, u is treated as a distribution, and the convolution acts as a smoothing or feature extraction operation.

In the same way, Dirac's Delta $\delta(\mathbf{x} - \mathbf{x}_0)$ can be seen as a perfect sensor such that captures the intensity at a specific point $\mathbf{x}_0$. In this sense, the delta function represents a very localized test function such that $\langle u, \delta_{\mathbf{x}_0} \rangle = u(\mathbf{x}_0.)$

For example, Figs. 3.3 and 3.4 give a suggestive representation for Heaviside function H and Dirac's Delta δ.

Fig. 3.3 The Heaviside function as an image

3.6 Derivatives of Distributions

Fig. 3.4 An image metaphor for Dirac's delta δ

3.6 Derivatives of Distributions

A distributional derivative extends the concept of weak derivatives even further, applying to distributions (generalized functions) rather than just integrable functions. The idea is to generalize the notion of the derivative to functions that may not be classically differentiable. This framework allows us to work with weak solutions to differential equations, where solutions are interpreted in the sense of distributions rather than classical functions. This is particularly useful in variational problems, where we deal with functionals involving derivatives of non-smooth functions.

The distributional derivative extends the classical derivative to a broader class of functions, facilitating the analysis of variational problems and differential equations in a more general and flexible setting. By using integration by parts, we can define the distributional derivative in a way that is consistent with classical derivatives when they exist, but also applicable to more singular or irregular functions.

We define the distributional derivative u' of a function $u \in \mathcal{L}^1_{loc}(\Omega)$ as the linear functional that acts on test functions $\varphi \in C_0^\infty$ by:

$$\langle u', \varphi \rangle = -\langle u, \varphi' \rangle$$

Here, $\langle u', \varphi \rangle$ denotes the action of the distribution u on the test function φ, which is typically the integral of u against φ

$$\langle u, \varphi \rangle = \int_\Omega u\varphi$$

In the same way the distribution $\frac{\partial u}{\partial x_k}$ is defined by

$$\langle \frac{\partial u}{\partial x_k}, \varphi \rangle = -\langle u, \frac{\partial \varphi}{\partial x_k} \rangle$$

$$= -\int u(\mathbf{x}) \frac{\partial}{\partial x_k} \varphi(\mathbf{x}) d\mathbf{x}$$

so we are led to define derivatives for a general distribution T in the same way:

$$\langle \frac{\partial T}{\partial x_k}, \varphi \rangle = -\langle T, \frac{\partial \varphi}{\partial x_k} \rangle \qquad (3.16)$$

which also defines a linear functional on φ, that is, if $\varphi_j \to \varphi$ in $\mathcal{D}$, then $\partial_k \varphi_j \to \partial_k \varphi$ and following with this kind of reasoning $\langle \partial_k T, \varphi_j \rangle$ converges to $\langle \partial_k T, \varphi \rangle$.

Next, second order derivatives should be considered. In classical calculus, a function $u : \mathbb{R}^N \to$ is said to have mixed partial derivatives if the order of differentiation does not matter, that is $\frac{\partial^2 u}{\partial x_j \partial x_i} = \frac{\partial^2 u}{\partial x_i \partial x_j}$. This result is guaranteed under certain smoothness conditions, typically if u is at least twice continuously differentiable ($u \in C^2$).

However, in the theory of distributions, these smoothness requirements are relaxed. Distributions allow us to handle functions or objects that may not be differentiable in the classical sense, yet still provide meaningful interpretations of their derivatives. This can be seen

$$\langle \frac{\partial^2 T}{\partial x_i \partial x_j}, \varphi \rangle = -\langle \frac{\partial T}{\partial x_j}, \frac{\partial \varphi}{\partial x_i} \rangle = \langle T, \frac{\partial^2 \varphi}{\partial x_j \partial x_i} \rangle$$

and by similar calculations

$$\langle \frac{\partial^2 T}{\partial x_j \partial x_i}, \varphi \rangle = \langle T, \frac{\partial^2 \varphi}{\partial x_i \partial x_j} \rangle$$

Given that φ has continuous second-order derivatives, it is well known that $\frac{\partial^2 \varphi}{\partial x_j \partial x_i} = \frac{\partial^2 \varphi}{\partial x_i \partial x_j}$ and finally we have that

$$\langle \frac{\partial^2 T}{\partial x_i \partial x_j}, \varphi \rangle = \langle \frac{\partial^2 T}{\partial x_j \partial x_i}, \varphi \rangle$$

For any multi-index $\alpha = (\alpha_1, \cdots, \alpha_n)$, we may continue this process and obtain

$$\langle D^\alpha T, \varphi \rangle = (-1)^{|\alpha|} \langle T, D^\alpha \varphi \rangle$$

So we may conclude that any distribution T has derivatives of all orders and the order of differentiation may be changed.

Generalizing a bit more, let $\mathcal{P}$ be a differential operator with constant coefficients

$$\mathcal{P} = \sum_\alpha c_\alpha D^\alpha$$

the transpose or adjoint operator is

$$\mathcal{P}^* = \sum_\alpha (-1)^{|\alpha|} c_\alpha D^\alpha$$

then

$$\langle \mathcal{P} T, \varphi \rangle = \langle T, \mathcal{P}^* \varphi \rangle$$

3.6 Derivatives of Distributions

Example Suppose u is a piecewise smooth function on $\mathbb{R}$ that is differentiable at all $x \neq 0$ but has a jump discontinuity at 0. We want to calculate its distributional derivative Du. By (3.16)

$$\langle Du, \varphi \rangle = -\langle u, \varphi' \rangle = -\int u\varphi' = -\int_{-\infty}^{0} - \int_{0}^{+\infty} \,;$$

the first integral is

$$-\int_{0}^{\infty} u\varphi' = -[u\varphi]_{0}^{+\infty} + \int_{0}^{\infty} u'\varphi$$

$$= u(0^+)\varphi(0) + \int_{0}^{\infty} u'\varphi$$

Similarly,

$$-\int_{-\infty}^{0} u\varphi' = -u(0-)\varphi(0) + \int_{-\infty}^{0} u'\varphi$$

adding the two results

$$\langle Du, \varphi \rangle = \sigma\varphi(0) + \int u'\varphi = \langle \sigma\delta + u', \varphi \rangle$$

and finally

$$Du = u' + \sigma\delta$$

where the "jump" at $x = 0$ is $\sigma = f(0+) - f(0-)$ (Fig. 3.5). □

If u is piecewise smooth on $\mathbb{R}$ with discontinuities at $x_1, x_2, \cdots$ with jumps σ_k and its pointwise derivative u' exists and is continuous except at the x_k's and perhaps some points where it has jump discontinuities, then

$$Du = u' + \sum \sigma_k \delta(x - x_k)$$

Fig. 3.5 Derivative for a jump function $u(x)$

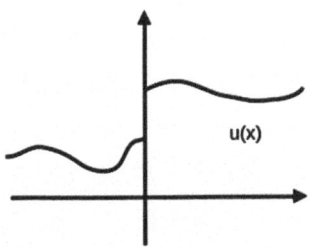

Sobolev Spaces

Sobolev spaces $H^m(\Omega)$ are a class of function spaces that are fundamental in the study of partial differential equations (PDEs) and variational problems. They extend the concept of differentiability to a broader, more flexible context, allowing for the analysis of functions whose derivatives may not be classically defined but can be understood in a weaker (distributional) sense. Variational methods in image processing, such as denoising and segmentation, often use Sobolev spaces to define smoothness constraints on images.

The Sobolev space $H^m(\Omega)$ is defined as the space of functions u such that: $u, D^\alpha u \in \mathcal{L}^2(\Omega)$ for all multi-indices $|\alpha| \leq m$, here, D^α denotes the weak derivative corresponding to the multi-index α.

$H^m(\Omega)$ is a Hilbert space with the inner product

$$\langle u, v \rangle_{H^m} = \int_\Omega \sum_{|\alpha| \leq m} (D^\alpha u)(D^\alpha v) d\mathbf{x} \qquad \forall u, v \in H^m$$

$$= \sum_{|\alpha| \leq k} \langle D^\alpha u, D^\alpha v \rangle_{\mathcal{L}^2}$$

that generate the Sobolev norm $\|\cdot\|_{H^m}$, with

$$\|u\|_{H^m}^2 = \sum_{|\alpha| \leq m} \|D^\alpha u\|_{\mathcal{L}^2}^2 \tag{3.17}$$

which written out in detail for the case $m = 2$ and $\Omega \subset \mathbb{R}^2$, becomes

$$\|u\|_{H^2}^2 = \int_\Omega (u^2 + u_x^2 + u_y^2 + u_{xx}^2 + 2u_{xy}^2 + u_{yy}^2) d\mathbf{x}$$

Some Sobolev spaces are

- $H^0(\Omega) = \mathcal{L}^2(\Omega)$, the space of square-integrable functions
- $H^1(\Omega)$ This space includes functions whose first weak derivatives are also in $\mathcal{L}^2(\Omega)$
- $H^2(\Omega)$ includes functions whose second weak derivatives are in $\mathcal{L}^2(\Omega)$.

3.7 Fourier Analysis

Although Fourier analysis is not inherently a variational method, it serves as a powerful and versatile tool for addressing a wide range of computational challenges

3.7 Fourier Analysis

in image processing. Even more, by leveraging the Fourier transform, variational regularization problems can often be reformulated into a more tractable form.

Definition 3.5 (Fourier Series) Let f be a periodic function with period 2π and integrable over $[-\pi, \pi]$ the Fourier series of f is the representation

$$f \sim \sum_{-\infty}^{\infty} c_k e^{ikx} = \frac{1}{2}a_0 + \sum_{k=1}^{\infty}(a_k \cos kx + b_k \sin kx) \qquad (3.18)$$

with

$$c_k = \frac{1}{2\pi} \int_{-\pi}^{\pi} f(x) e^{-ikx} dx \qquad (3.19)$$

Definition 3.6 (Periodic Convolution) Given two 2π-periodic integrable functions f, u on $\mathbb{R}$ their **convolution** $f * u$ on $[-\pi, \pi]$ is

$$f * u(x) := \frac{1}{2\pi} \int_{-\pi}^{\pi} f(x-t)u(t) dt \qquad (3.20)$$

One of the most useful relations between Fourier series and convolution is the following

$$S_N^f(x) := \sum_{k=-N}^{N} c_k e^{ikx}$$

$$= \sum_{k=-N}^{N} \left(\frac{1}{2\pi} \int_{-\pi}^{\pi} f(t) e^{-ikt} dt \right) e^{ikx}$$

$$= \frac{1}{2\pi} \int_{-\pi}^{\pi} f(t) \left(\sum_{-N}^{N} e^{ik(x-t)} \right) dt$$

$$= (f * D_N)(x)$$

where D_N is the Nth Dirichlet **Kernel** (Fig. 3.6) given by

$$D_N(x) = \sum_{k=-N}^{N} e^{ikx}$$

It can be shown [43] that

$$D_N(x) = \frac{\sin(N + 1/2)x}{\sin \frac{1}{2}x} \qquad (3.21)$$

Fig. 3.6 Dirichlet kernel

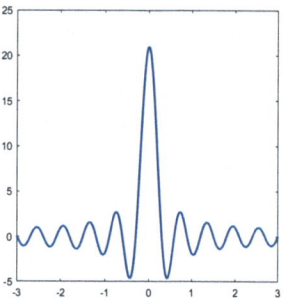

So the problem of understanding Fourier series can be reduced to study the convolution $f * D_N$. In general, a family of kernels $\{K_n(x)\}_{n=1}^{\infty}$ on $I =]-a, a[$ is said to be a family of **good kernels** or **positive summation kernel** if it satisfies the following properties:

(a) For all $n \geq 1$

$$K_n(t) \geq 0.$$

(b) The integral

$$\int_{-a}^{a} Kn(t)dt = 1$$

(c) For every $\delta > 0$

$$\int_{\delta \leq |t| \leq a} |K_n(t)|dt \to 0, \qquad n \to \infty$$

These properties ensure that when the kernel is used in convolution operations, it acts as a smoothing operator without altering the overall scale of the signal.

Theorem 3.1 *With the above conditions, if $f : I \to \mathbb{R}$ is integrable and bounded on I continuous for $t = 0$*

$$\lim_{n \to \infty} \int_{-a}^{a} K_n(t)f(t)dt = f(0)$$

Example The sequence defined by $K_n : \mathbb{R} \to \mathbb{R}$ such that

$$K_n(x) = \begin{cases} n, & |x| < \frac{1}{2n} \\ 0, & |x| > \frac{1}{2n} \end{cases}$$

is a good kernel.

3.8 Convolution and Fourier Transform in $\mathbb{R}$

Fig. 3.7 A family of good kernels $K_n(x) = n\psi(nx)$ as in (3.22)

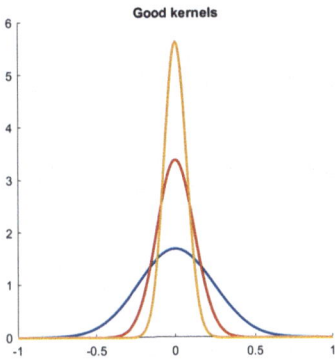

Example If $\psi(x) = \frac{1}{\sqrt{2\pi}} e^{-\frac{x^2}{2}}$ for the family defined by

$$K_n(x) = n\psi(nx) \qquad (3.22)$$

the three above conditions are fulfilled and by consequence is a good kernel on $\mathbb{R}$ (Fig. 3.7).

□

As a corollary if $\{K_n\}_{n=1}^{\infty}$ is a good kernel on the interval I, t_0 an interior point of I, and f continuous at $t = t_0$, then

$$\lim_{n \to \infty} \int_I K_n(t) f(t_0 - t) dt = f(t_0)$$

Theorem 3.2 *Let $\{K_n\}_{n=1}^{\infty}$ be a family of good kernels, and f an integrable function on the circle. Then*

$$(f * K_n)(x) \to f(x) \qquad n \to \infty \qquad (3.23)$$

Note that positive summation kernels can be seen as approximations to Dirac's δ. As the width of a kernel decreases and its height increases such that the area under the curve remains 1, the kernel approaches the behavior of δ function.

3.8 Convolution and Fourier Transform in $\mathbb{R}$

Convolution is an integral operation that combines two functions to produce a third function, often used to apply filters to images or signals. Convolution has a solid mathematical foundation and it is a linear operation. This makes it easier to analyze and design complex image processing pipelines.

Convolutions and Fourier transforms can be defined in $\mathcal{L}^1(\mathbb{R})$ or $\mathcal{L}^2(\mathbb{R})$. One space is not more important than the other, $\mathcal{L}^1$ is not subset of $\mathcal{L}^2$ and vice versa. Both spaces have their specific advantages and disadvantages, and the choice of space often depends on the particular application and the properties of the functions being analyzed [112].

Definition 3.7 If f, u are functions on $\mathbb{R}$ their convolution is the function $f * u$ defined by

$$f * u(x) := \int_{-\infty}^{\infty} f(x - t)u(t)dt \qquad (3.24)$$

provided that the integral exists. There are many properties that can be required on the functions f and u so that the convolution is well defined. From now on we assume that f and u have the necessary properties for the integrals to be absolutely convergent $\forall x \in \mathbb{R}$.

Assuming all the terms exist, convolution behaves as ordinary multiplication:

- $f * (\alpha u + \beta v) = \alpha(f * u) + \beta(f * v)$
- $f * u = u * f$
- $f * (u * v) = (f * u) * v$

This is seen from the definition. For example, doing the change of variable $\xi = x - t$

$$f * u(x) = \int f(x - t)u(t)dt = \int f(\xi)u(x - \xi)d\xi = u * f(x)$$

Convolutions are easily differentiable, given that

$$(f * u)'(x) = \frac{d}{dx}\int f(x - t)u(t)dt = \int f'(x - t)u(t)dt = (f' * u)(x)$$

and by commutative property $(f * u)' = f' * u = u' * f$. So the convolution $f * u$ inherits the smoothness properties of f and u.

Note that

$$\int f(x - t)u(t)dt \approx \sum f(x - t_k)u(t_k)\Delta t_k$$

so, the sum is a linear combination of translates of f with coefficients $u(t_k)\Delta t_k$, then $f * u$ can be considered as a continuous superposition of translates. By this property, convolution is applied in the **smoothing** of signals. For example, if

$$u(x) = \begin{cases} \frac{1}{2\varepsilon}, & -\varepsilon \leq x \leq \varepsilon \\ 0, & \text{otherwise} \end{cases}$$

3.8 Convolution and Fourier Transform in $\mathbb{R}$

then

$$f * u(x) = \frac{1}{2\varepsilon} \int_{x-\varepsilon}^{x+\varepsilon} f(t) dt$$

which is the average of f in the interval $[x-\varepsilon, x+\varepsilon]$. Smoothing an image involves averaging the pixel values in a local neighborhood. This helps reduce noise and small variations in pixel intensity, resulting in a smoother image.

Direct Product and Convolution

Remember a distribution is defined by its action on a function $\varphi \in C_0^\infty$. Then restricting to the one variable case the multiplication of two functions f, u locally integrable in $\mathbb{R}$ can be seen as

$$\langle f(x)u(y), \varphi(x,y)\rangle = \int f(x) \int u(y)\varphi(x,y) dy dx$$
$$= \langle f(x), \langle u(y), \varphi(x,y)\rangle\rangle$$

This reasoning suggests that if T, S are distributions the direct product $T \otimes S$ can be defined as

$$\langle T \otimes S, \varphi(x,y)\rangle := \langle T, \langle S, \varphi(x,y)\rangle\rangle$$

Using the former definition we can define Convolution in the sense of distributions:

$$\langle f * u, \varphi \rangle = \int (f * u)(z)\varphi(z) dz = \int \left[\int u(y) f(z-y) dy\right] \varphi(z) dz$$
$$= \int u(y) \left[\int f(z-y)\varphi(z) dz\right] dy = \int u(y) \left[\int f(x)\varphi(x+y) dx\right] dy$$
$$= \int f(x) u(y) \varphi(x+y) dx dy = \langle f(x) \otimes u(y), \varphi(x+y)\rangle, \varphi \in \mathcal{D}$$

Then, for distributions S, T convolution is defined as

$$\langle T * S, \varphi \rangle = \langle T \otimes S, \varphi(x+y)\rangle \tag{3.25}$$

For ordinary functions we know the shift or translation $\tau_a f = f(x-a)$, in terms of distributions

$$\langle f(x-a), \varphi \rangle = \int_{-\infty}^{+\infty} f(x-a)\varphi(x) dx = \int f(x)\varphi(\xi+a) d\xi = \langle f, \varphi(x+a)\rangle$$

Then the definition for a distribution T is

$$\langle \tau_a T, \varphi \rangle := \langle T, \varphi(x+a) \rangle$$

An important property of the δ with convolutions is $\delta_{(a)} * T = \tau_a T$ because

$$\langle \delta_{(a)} * T, \varphi \rangle = \langle T_y, \langle \delta_{(a)_x}, \varphi(x+y) \rangle \rangle = \langle T_y, \varphi(a+y) \rangle = \langle \tau_a T, \varphi \rangle$$

then as a consequence we have

$$\delta * T = T; \qquad \delta_{(a)} * f = f(x-a) \tag{3.26}$$

Theorem 3.3 ([43]) *Let u be an $\mathcal{L}^1(\mathbb{R})$ function such that $\int_{-\infty}^{+\infty} u(x)dx = 1$, and let $\alpha = \int_{-\infty}^{0} u(x)dx$ and $\beta = \int_{0}^{+\infty} u(x)dx$ (Note that $\alpha + \beta = 1$ Suppose that f is piecewise continuous on $\mathbb{R}$, and suppose either that f is bounded or that u vanishes outside a finite interval so that $f * u(x)$ is well-defined for all x. If $u(t)$ is defined by*

$$u_\varepsilon(x) = \frac{1}{\varepsilon} u\left(\frac{x}{\varepsilon}\right) \tag{3.27}$$

then

$$\lim_{\varepsilon \to 0} f * u_\varepsilon(x) = \alpha f(x+) + \beta f(x-)$$

for all x.

A well-known variant of the previous property is the following

Theorem 3.4 *Suppose $u \in \mathcal{L}^1(\mathbb{R})$ is bounded and satisfies $\int u(t)dt = 1$. If $f \in \mathcal{L}^2$ then $f * u(x)$ is well-defined for all x, and*

$$f * u_\varepsilon \to f$$

converges to f in norm as $\varepsilon \to 0$.

The Fourier Transform in $\mathbb{R}$

If f is an integrable function on $\mathbb{R}$ its Fourier Transform is the function $\widehat{f}$ on $\mathbb{R}$ defined by

$$\widehat{f}(\xi) := \int e^{-i\xi x} f(x) dx \tag{3.28}$$

3.8 Convolution and Fourier Transform in $\mathbb{R}$

It is also used the symbol

$$\mathcal{F}[f(x)] = \widehat{f}(\xi)$$

Since $|e^{-i\xi x}| = 1$, this integral converge for all ξ and is bounded by

$$|\widehat{f}(\xi)| \leq \int |f(x)|dx.$$

Properties of Fourier Transform

Suppose $f \in \mathcal{L}^1$, that is, $\int |f| < \infty$. The Fourier transform is a linear operator:

$$\mathcal{F}(f+g) = \mathcal{F}(f) + \mathcal{F}(g); \qquad \mathcal{F}(\alpha f) = \alpha \mathcal{F}(f)$$

1. *Shift Theorem.* For $a \in \mathbb{R}$

$$\mathcal{F}[f(x-a)] = e^{-ia\xi}\widehat{f}(\xi), \text{ and} \qquad \mathcal{F}[e^{iax}f(x)] = \widehat{f}(\xi - a) \qquad (3.29)$$

2. If $\delta > 0$ and $f_\delta(x) := \dfrac{1}{\delta} f(\dfrac{x}{\delta})$, then

$$[\widehat{f_\delta}](\xi) = \widehat{f}(\delta\xi), \text{ and } \mathcal{F}[f(\delta x)] = [\widehat{f}]_\delta(\xi)$$

3. If f is continuous and piecewise and $f' \in \mathcal{L}^1$, then

$$\widehat{f^{(k)}}(\xi) = (i\xi)^k \widehat{f}(\xi) \qquad (3.30)$$

4. If also $u \in \mathcal{L}^1$, then (*Convolution Theorem*)

$$\widehat{f*u}(\xi) = \widehat{f}(\xi) \cdot \widehat{u}(\xi) \qquad (3.31)$$

This last property is given by

$$\widehat{f*u} = \iint e^{-i\xi x} f(x-y)u(y)dydx$$

$$= \iint e^{-i\xi(x-y)} f(x-y) e^{-i\xi y} u(y) dxdy$$

$$= \iint e^{-i\xi z} f(z) e^{-i\xi y} u(y) dzdy$$

$$= \widehat{f}(\xi)\widehat{u}(\xi)$$

5. Parseval's Identity $2\pi \langle f, u \rangle = \langle \widehat{f}, \widehat{u} \rangle$, given by

$$2\pi \langle f, u \rangle = \int f(x)\overline{u(x)}dx = \iint f(x)\overline{e^{i\xi x}\widehat{u}(\xi)}d\xi dx$$
$$= \iint f(x)e^{-i\xi x}\overline{\widehat{u}(\xi)}dx d\xi = \int \widehat{f}(\xi)\overline{\widehat{u}(\xi)}d\xi$$
$$= \langle \widehat{f}, \widehat{u} \rangle.$$

and
6.

$$\int |f(x)|^2 dx = \frac{1}{2\pi}\int |\widehat{f}(\xi)|^2 d\xi \qquad \textit{Plancherel formula} \qquad (3.32)$$

Fourier Transform of Distributions

Such as we have done before, when working with generalized functions, the Fourier transform is defined not directly by a formula, but through its action on test functions. Instead of evaluating an integral, we define the Fourier transform by how it interacts with regular (well-behaved) test functions.

Let T be a distribution and let φ be a test function from a space of well-behaved functions (called the *Schwartz space*, see [108]). The Fourier transform $\widehat{T}$, is defined by

$$\langle \widehat{T}, \varphi \rangle = \langle T, \widehat{\varphi} \rangle$$

Example (Fourier Transform of δ) Remember $\widehat{u}(\xi) = \int u(x)e^{-i\xi x}dx$, then

$$\langle \widehat{\delta}, \varphi \rangle = \langle \delta, \widehat{\varphi} \rangle = \int \delta e^{-i\xi x}$$
$$= \langle \delta, e^{-i\xi x} \rangle = e^{-i(0)\xi}$$
$$= 1$$

so

$$\widehat{\delta}(\xi) = 1 \qquad (3.33)$$

is the Fourier Transform of Dirac's Delta.

3.9 Convolution and Fourier Transforms in $\mathbb{R}^N$

Most ideas and definitions about convolution and Fourier transform given above can be generalized to several variables. Then convolution is defined as

$$(f * u)(\mathbf{x}) = \int f(\mathbf{x} - t)u(t)dt$$

Definition 3.8 For $\mathbf{y} \in \mathbb{R}^N$ the *shift of* f by $\mathbf{y}$ is the function $\tau_\mathbf{y} f$ defined by $\tau_\mathbf{y} f(\mathbf{x}) = f(\mathbf{x} - \mathbf{y})$

An operator $\mathscr{A}$ is shift invariant if

$$\mathscr{A}(\tau_\mathbf{y} f) = \tau_\mathbf{y}(\mathscr{A} f)$$

Theorem 3.5 *An operator defined by convolution is shift invariant.*
If $\mathscr{A}(f) := f * u$ *then*

$$\mathscr{A}(\tau_\mathbf{y} f)) = \int f(\mathbf{x} - \mathbf{y} - t)u(t)dt$$

$$= (f * u)(\mathbf{x} - \mathbf{y})$$

$$= \tau_\mathbf{y}(\mathscr{A} f)$$

Other properties of convolution are the following

$$\frac{\partial}{\partial x_k} f * u(\mathbf{x}) = \frac{\partial f}{\partial x_k} * u(\mathbf{x}) = f * \frac{\partial u}{\partial x_k}(\mathbf{x}) \qquad (3.34)$$

This is because

$$\frac{\partial}{\partial x_k} f * u(\mathbf{x}) = \frac{\partial}{\partial x_k} \int f(\mathbf{x} - \xi)u(\xi)d\xi$$

$$= \int \frac{\partial}{\partial x_k} f(x - \xi)u(\xi)d\xi$$

$$= \frac{\partial f}{\partial x_k} * u(\mathbf{x})$$

In the language of distributions

$$\langle \frac{\partial}{\partial x_k} T * S, \varphi \rangle = (-1)\langle T * S, \frac{\partial \varphi}{\partial x_k} \rangle = -\langle T \otimes S, \frac{\partial \varphi}{\partial x_k}(x + y) \rangle$$

$$= -\langle T_x, \langle S_y, \frac{\partial \varphi}{\partial x_k}(x + y) \rangle \rangle = \langle T, \langle \frac{\partial S}{\partial x_k}, \varphi(x + y) \rangle \rangle$$

$$= \langle T * \frac{\partial S}{\partial x_k}, \varphi \rangle$$

then
$$\frac{\partial}{\partial x_k} T * S = T * \frac{\partial S}{\partial x_k} \tag{3.35}$$

In this way, if $\mathcal{P}$ is a linear differential operator we have
$$\mathcal{P}(T * S) = T * (\mathcal{P}S) = (\mathcal{P}T) * S.$$

In particular
$$\mathcal{P}(\delta * T) = (\mathcal{P}\delta) * T = \delta * (\mathcal{P}T) = \mathcal{P}(T) \tag{3.36}$$

Example (Fundamental Solutions) Given a linear differential operator with constant coefficients
$$\mathcal{P} = \sum_{|\alpha| \leq m} c_\alpha \partial^\alpha,$$

the distribution $K(\mathbf{x}, \mathbf{y})$ is a fundamental solution of $\mathcal{P}$ if
$$\mathcal{P}_\mathbf{x}(K(\mathbf{x}, \mathbf{y})) = \delta(\mathbf{x} - \mathbf{y}) \tag{3.37}$$

Here $\mathcal{P}$ is applied to K as a function of $\mathbf{x}$ and $\mathbf{y}$ is a parameter. For example, if $\Delta K = \delta$ then K is called a potential. Fundamental solutions have the remarkable property of being part of the solution of inhomogeneous differential equations. Reasoning formally and using (3.36), if $\mathcal{P}K = \delta$ then
$$\mathcal{P}(K * f) = (\mathcal{P}K) * f = \delta * f = f$$

therefore
$$u = K * f \quad \text{is a solution of} \quad \mathcal{P}u = f \tag{3.38}$$

Example Let us find the fundamental solution for the 4th derivative (biharmonic) operator $\mathcal{P}u = u^{(iv)}$, that is, the solution of the differential equation
$$u^{(iv)} = \delta$$

Taking the Fourier transform of both sides (3.30), (3.33):
$$\widehat{u^{(iv)}} = \widehat{\delta}; \rightarrow \xi^4 \widehat{u} = 1; \rightarrow \widehat{u} = \frac{1}{\xi^4}$$

3.9 Convolution and Fourier Transforms in $\mathbb{R}^N$

then

$$u = \mathscr{F}^{-1}\left[\frac{1}{\xi^4}\right]$$

This expression can be evaluated using known Fourier transform pairs. The inverse Fourier transform of $\frac{1}{\xi^4}$ is proportional to the Green's function of the biharmonic operator D^4, which is a *cubic spline kernel*. Specifically, for $\frac{1}{\xi^4}$, the inverse Fourier transform is:

$$u(x) = \frac{|x|^3}{6} \tag{3.39}$$

As for the solution of the equation

$$u^{(iv)} = \delta_a \tag{3.40}$$

applying the same methods

$$u = \mathscr{F}^{-1}\left[\frac{e^{-ia\xi}}{\xi^4}\right]$$

so, by properties (3.29), we recognize that this is simply a shifted version of the solution for $u^{(iv)} = \delta$, from the previous solution (3.39) and applying the shift a

$$u(x) = \frac{|x-a|^3}{6} \tag{3.41}$$

This is a cubic polynomial spline centered at $x = a$, which represents the fundamental solution of the biharmonic operator D^4 with a delta function shifted by a

Example A generalization of the previous example are the fundamental solutions of the iterated Laplacian operator $\Delta^m u$ in $\mathbb{R}^d$. If

$$\Delta^m K = \delta$$

then the fundamental solution K is (see [122])

$$K(\mathbf{x}) = \begin{cases} \alpha_{m,d}\, \|\mathbf{x}\|^{2m-d} \ln \|\mathbf{x}\|, & (2m-d) \text{ even integer} \\ \alpha_{m,d}\, \|\mathbf{x}\|^{2m-d}, & \text{otherwise,} \end{cases} \tag{3.42}$$

where

$$\alpha_{m,d} = \begin{cases} \frac{(-1)^{d/2+1+m}}{2^{2m-1}\pi^{d/2}(m-1)!(m-d/2)!}, & (2m-d) \text{ even integer} \\ \frac{\Gamma(d/2-m)}{2^{2m}\pi^{d/2}(m-1)!}, & \text{otherwise.} \end{cases}$$

In practical applications, we don't have to worry about this formula for $\alpha_{m,d}$ given that coefficients are assimilated by scalars in linear combinations. □

Convolution Discretization

In order to be useful for data science, convolution can be discretized by taking

$$f * u = \int f(x-t)u(t)dt \approx \sum f(x-t_j)u(t_j)\Delta t$$

If we think about f, u as the discrete values in the vectors $[f(j)]_{j=1}^m$ and $[u(j)]_{j=1}^n$ (Fig. 3.8), then the discrete convolution $h = f * u$ is

$$h(k) = \sum f(j)u(k-j+1)$$

Convolutions can be generalized to higher dimensions, for the problems of Image processing corresponds to a two dimensional convolution

$$f * u(x, y) = \int_{-\infty}^{+\infty} \int_{-\infty}^{+\infty} f(t_1, t_2) u(x-t_1, y-t_2) dt_1 \, dt_2$$

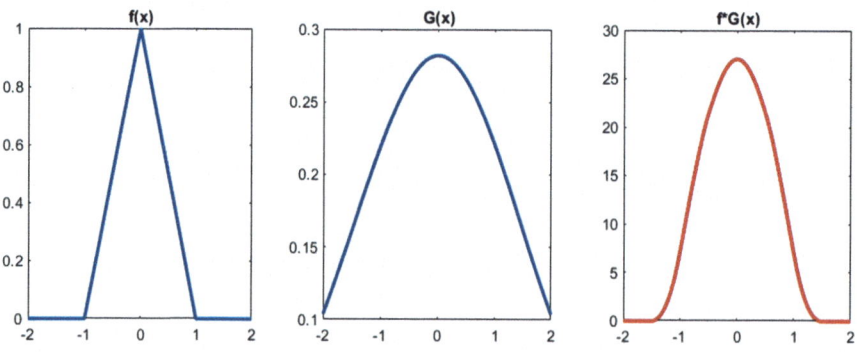

Fig. 3.8 One dimensional convolution example. $G(x) = \frac{1}{\varepsilon} u(\frac{x}{\varepsilon})$ as in (3.27)

3.9 Convolution and Fourier Transforms in $\mathbb{R}^N$

Fig. 3.9 Comparing smoothing between median filtering and a basic convolution function from ChatGpt

Lena

Salt & pepper noise

Median filtering

Convolution Smoothing

with discrete form

$$c(j,k) = \sum_p \sum_q f(p,q) u(j-p+1, k-q+1)$$

where f, u are matrices and p and q run over all values that lead to legal subscripts of $f(p,q)$ and $u(j-p+1, k-q+1)$.

Example (2D Convolution) Figure 3.9 shows the result of reducing the noise in an image by convolution. In one case is applied a convolution filter with the matlab function *medfilt2* and the other with kernel

$$u = \begin{bmatrix} 1 & 1 & 1 \\ 1 & 1 & 1 \\ 1 & 1 & 1 \end{bmatrix} \quad (3.43)$$

Fourier transform definition in $\mathbb{R}^N$ is an elegant generalization of the one-dimensional case. If $f \in \mathcal{L}^1(\mathbb{R}^N)$, its Fourier transform $\widehat{f}$ is a bounded function on $\mathbb{R}^N$ defined by

$$\widehat{f}(\xi) := \int_{\mathbb{R}^N} e^{-i\mathbf{x}\cdot\boldsymbol{\xi}} f(\mathbf{x}) d\mathbf{x} \quad (3.44)$$

where $\mathbf{x}, \boldsymbol{\xi} \in \mathbb{R}^N$, $\mathbf{x}\cdot\boldsymbol{\xi} = x_1\xi_1 + \cdots + x_N\xi_N$

1. Translation:

$$\text{If } \tau_{\mathbf{y}} f(\mathbf{x}) = f(\mathbf{x} - \mathbf{y}) \quad \text{change of variable } \mathbf{x} \to \mathbf{x} + \mathbf{y}$$

$$\widehat{\tau_{\mathbf{y}} f}(\boldsymbol{\xi}) = \int_{\mathbb{R}^N} e^{-i\mathbf{x}\cdot\boldsymbol{\xi}} f(\mathbf{x}-\mathbf{y}) d\mathbf{x}$$

$$= \int_{\mathbb{R}^N} e^{-i(\mathbf{x}+\mathbf{y})\cdot\boldsymbol{\xi}} f(\mathbf{x}) d\mathbf{x}$$

$$= e^{-i\mathbf{y}\cdot\boldsymbol{\xi}} \widehat{f}(\boldsymbol{\xi})$$

2. As we could expect, multiplication by an exponential has a dual of the previous property with translation of Fourier Transform

$$\mathscr{F}(e^{i\mathbf{x}\cdot\mathbf{y}} f(\mathbf{x}))(\boldsymbol{\xi}) = \int e^{-i\mathbf{x}\cdot\boldsymbol{\xi}} e^{i\mathbf{x}\cdot\mathbf{y}} f(\mathbf{x}) d\mathbf{x}$$

$$= \int e^{-i\mathbf{x}\cdot(\boldsymbol{\xi}-\mathbf{y})} f(\mathbf{x}) d\mathbf{x}$$

$$= \widehat{f}(\boldsymbol{\xi} - \mathbf{y}) = \tau_{\mathbf{y}} \widehat{f}(\boldsymbol{\xi})$$

3. Differentiation:

$$\widehat{\frac{\partial f}{\partial x_k}}(\boldsymbol{\xi}) = -i\xi_k \widehat{f}(\boldsymbol{\xi}) \tag{3.45}$$

$$\widehat{\frac{\partial f}{\partial x_k}}(\boldsymbol{\xi}) = \int e^{-i\mathbf{x}\cdot\boldsymbol{\xi}} \frac{\partial f}{\partial x_k}(\mathbf{x}) d\mathbf{x}$$

$$= -\int f(\mathbf{x}) \frac{\partial}{\partial x_k}(e^{-i\mathbf{x}\cdot\boldsymbol{\xi}}) d\mathbf{x}$$

$$= i\xi_k \int e^{-i\mathbf{x}\cdot\boldsymbol{\xi}} f(\mathbf{x}) d\mathbf{x}$$

$$= i\xi_k \widehat{f}(\boldsymbol{\xi})$$

Definition 3.9 (Inverse Fourier Transform) For a function f for which $\int_{\mathbb{R}^N} |f(\boldsymbol{\xi})| d\boldsymbol{\xi} < \infty$, the inverse Fourier Transform $\mathscr{F}^{-1}$ of f is defined, for each $\mathbf{x} \in \mathbb{R}^N$ (Fig. 3.10), by

$$(\mathscr{F}^{-1}\mathscr{F}f)(\mathbf{x}) := \frac{1}{(2\pi)^N} \int \widehat{f}(\boldsymbol{\xi}) e^{i\mathbf{x}\cdot\boldsymbol{\xi}} d\boldsymbol{\xi}$$

3.9 Convolution and Fourier Transforms in $\mathbb{R}^N$

Original

$f * u = \mathcal{F}^{-1}(\hat{f} \cdot \hat{u})$

Fig. 3.10 Original noisy image and its smoothing by convolution using inverse Fourier Transform $f * u = \mathcal{F}^{-1}(\hat{f} \cdot \hat{u})$

and the inversion formula is

$$\mathcal{F}^{-1}(\mathcal{F}f)(\mathbf{x}) = f(\mathbf{x})$$

Example Observe from (3.31) that $\widehat{f * u} = \hat{f}\hat{u}$, implies

$$f * u = \mathcal{F}^{-1}(\hat{f} \cdot \hat{u}) \qquad (3.46)$$

which can be used calculating convolution with Fourier transform. The following matlab code is a basic implementation (from [84]) of this formula. In this case we are convolving the image f with the kernel (3.43)

```
k2=[1 1 1;1 1 1;1 1 1];
f=imread('cameraman.jpg');
f=double(f(:,:,1));
FT=fft2(f);
TransfK=fft2(pad(f, k2));
invertedT=ifft2(rearrange(FT.*TransfK));
subplot(1,2,1),imshow(f,[]),title('Original')
subplot(1,2,2),imshow(invertedT,[])
```

A well-known way to filter an image u is by $K_\sigma * u(x)$, where K_σ is the Gaussian kernel with standard deviation $\sigma > 0$

$$K_\sigma(\mathbf{x}) = \frac{1}{2\pi\sigma^2} e^{-\frac{\|\mathbf{x}\|^2}{2\sigma^2}} \qquad (3.47)$$

with $K_\sigma(x, y) = \frac{1}{2\pi\sigma^2} e^{\frac{-x^2-y^2}{2\sigma^2}}$ for the two-dimensional case.

$\square$

Finally, we have the analog in $\mathbb{R}^N$ for the dilation formula (3.27)

$$u_\varepsilon(\mathbf{x}) := \frac{1}{\varepsilon^N} u\left(\frac{\mathbf{x}}{\varepsilon}\right) \qquad (3.48)$$

and the corresponding N-dimensional version for one of the properties that justify our obsession for convolutions

If $u \in \mathcal{L}^1$ and $\int u(\mathbf{x}) d\mathbf{x} = 1$ and u_ε as in (3.48), the following holds:

- Suppose that either f is bounded or u vanishes outside a bounded set, so that $f * u$ is well defined. If f is continuous at $\mathbf{x}$, then

$$\lim_{\varepsilon \to 0} f * u_\varepsilon(\mathbf{x}) = f(\mathbf{x})$$

If f is continuous on a closed, bounded set Ω, the convergence is uniform on Ω

- if $f \in \mathcal{L}^2$, then

$$\lim_{\varepsilon \to 0} \|f * u_\varepsilon - f\| = 0$$

Matlab Code

Listing 3.1 2D-Convolution function

```
function output = conv2D(input, kernel)
% Input: input - input matrix
%        kernel - kernel matrix
% Output: output - output matrix after 2D convolution
%TAKEN FROM ChatGPT
input_rows = size(input, 1);
input_cols = size(input, 2);
kernel_size = size(kernel, 1);
pad_size = floor(kernel_size / 2);
% Assuming zero-padding to preserve output size

% Zero-pad the input matrix
input_padded = padarray(input, [pad_size, pad_size],0);

% Initialize the output matrix
output = zeros(input_rows, input_cols);

for i = 1:input_rows% Iterate over the input matrix
for j = 1:input_cols
for k = -pad_size:pad_size % Iterate over the kernel matrix
for l = -pad_size:pad_size
ii = i + k + pad_size;% Compute the input and kernel indices
jj = j + l + pad_size;
```

3.9 Convolution and Fourier Transforms in $\mathbb{R}^N$

```
kk = k + pad_size + 1;
ll = l + pad_size + 1;
% Update the output element
output(i, j) = output(i, j) + input_padded(ii, jj)...
* kernel(kk, ll);
end
end
end
end
```

Chapter 4
Minimization of Functionals

There is no doubt that contemporary differentiation (including Frechet derivative) had its origin in the calculus of variations. In this chapter, we delve into the foundational concepts of functional minimization within the framework of variational calculus, focusing on their applications in image processing. We begin by defining the Gâteaux derivative [114], a key tool for understanding variations in functionals, which paves the way for formulating optimization problems. Following this, we explore the first variation and the gradient of functionals, deriving the associated Euler-Lagrange equations for various fundamental functionals. Finally, we introduce variational methods in the context of minimal surface theory, where the surface area is minimized, offering deeper insights into both geometric problems and their practical applications in image processing. This chapter equips the reader with critical tools for formulating and solving variational problems [74, 98, 110].

4.1 The Gâteaux Variation

The Gâteaux differential is defined for a transformation (possibly nonlinear) $\mathcal{J}: U \subset \mathcal{U} \to \mathcal{V}$, with $\mathcal{U}$ a vector space and $\mathcal{V}$ a normed space. The space $\mathcal{U}$ does not need to be normed. This differential generalizes the concept of directional derivative usual in N-dimensional Euclidian space.

Definition 4.1 (Gâteaux Derivative) Let $\mathcal{J}$ be the transformation $\mathcal{J}: U \subset \mathcal{U} \to \mathcal{V}$, $u \in U$ and let φ be arbitrary in $\mathcal{U}$. Then the limit

$$\delta \mathcal{J}(u; \varphi) := \lim_{\varepsilon \to 0} \frac{1}{\varepsilon}[\mathcal{J}(u + \varepsilon\varphi) - \mathcal{J}(u)],$$

if exists, it is called the **Gâteaux Derivative** of $\mathcal{J}$ at u in the direction of φ.

If this limit exists for each $\varphi \in \mathcal{U}$, the transformation is called Gâteaux differentiable at u. If we fix $u \in U$ and consider φ as a variable, the Gâteaux derivative defines a transformation from $\mathcal{U}$ to $\mathcal{V}$.

The most frequent application of this definition is in the case where $\mathcal{V} = \mathbb{R}$ is the real line and hence the application $\mathcal{J}$ is a real-valued functional on $\mathcal{U}$. Thus, given a functional $\mathcal{J}$ in $\mathcal{U}$, the Gateaux derivative of $\mathcal{J}$ if it exists, is

$$\delta \mathcal{J}(u; \varphi) = \frac{\partial}{\partial \varepsilon} \mathcal{J}(u + \varepsilon \varphi) \bigg|_{\varepsilon = 0} \qquad (4.1)$$

and for each fixed $u \in \mathcal{U}$, $\delta \mathcal{J}(u; \varphi)$ is a linear functional with respect to the variable $\varphi \in \mathcal{U}$

Example Consider $\mathcal{U} = C[0, 1]$ and $\mathcal{J}(u) = \int_0^1 f(x, u(x)) dx$, then

$$\delta \mathcal{J}(u; \varphi) = \frac{\partial}{\partial \varepsilon} \int_0^1 f(x, u + \varepsilon \varphi) \, dx \bigg|_{\varepsilon = 0}$$

and assuming f_x exists and is continuous with respect to (u, x), then $\delta \mathcal{F}$ is the functional

$$\delta \mathcal{F}(u; \varphi) = \int_0^1 f_x(x, u(x)) \varphi(x) \, dx$$

Example If $\mathcal{U} = \mathbb{R}^N$ and $\mathcal{J} = f(\mathbf{x})$ is a functional on $\mathbb{R}^N$ with continuous partial derivatives, recall the directional derivative and the gradient of a function $f : \mathbb{R}^N \to \mathbb{R}$ are related by the directional derivative (2.11) which coincides with the Gateaux derivative of f

$$\delta f(\mathbf{x}; \mathbf{h}) = \sum_{j=1}^N \frac{\partial f}{\partial x_j} h_j = \langle \nabla f, \mathbf{h} \rangle$$

The same is true for infinite dimensional Banach Spaces □

The Gradient of a Functional

The notion of gradient can be extended to infinite dimensional spaces by considering functionals $\mathcal{J} : \mathcal{H} \to \mathbb{R}$ where $\mathcal{H}$ is an inner product space. Let's assume $\mathcal{J}$ is Gateaux differentiable at u. If there exists $\nabla \mathcal{J}(u) \in \mathcal{H}$ such that

$$\delta \mathcal{J}(u; \varphi) = \langle \nabla \mathcal{J}(u), \varphi \rangle = \frac{\partial}{\partial \varepsilon} \mathcal{J}(u + \varepsilon \varphi) \bigg|_{\varepsilon = 0} \qquad (4.2)$$

4.1 The Gâteaux Variation

Then $\nabla \mathcal{J}$ is the representer of the functional $\delta \mathcal{J}(u; \varphi)$ and is called the **gradient** or (**functional gradient**) of $\mathcal{J}$ at u. If the inner product space $\mathcal{H}$ is complete and as a consequence, a Hilbert Space, then by the Riesz representation theorem (3.4) the gradient $\nabla \mathcal{J}(u)$ in (4.2) always exists.

Applying these ideas to our classical problem (1.3). The quantity

$$\langle \nabla \mathcal{J}(y); \varphi \rangle = \int \left[\frac{\partial F}{\partial y} \varphi + \frac{\partial F}{\partial y'} \varphi' \right] dx \quad \forall \varphi \in C_0^\infty(a, b) \tag{4.3}$$

is called the **first variation** of $\mathcal{J}$. Commonly, in this and similar problems, an explicit expression for $\nabla \mathcal{J}(y)$ is obtained by writing the first variation in terms of the inner product as

$$\langle \nabla \mathcal{J}(y); \varphi \rangle = \int_a^b \nabla \mathcal{J}(y) \varphi \, dx = \int_a^b G \varphi \, dx \quad \forall \varphi \in C_0^\infty(a, b)$$

and from here conclude $G(y) = \nabla \mathcal{J}(y)$.

Using integration by parts we can replace the term with φ' in (4.3) and represent $\langle \nabla \mathcal{J}(y); \varphi \rangle$ as a linear functional on $C_0^\infty(a, b)$

$$\int_a^b \nabla \mathcal{J}(y) \varphi \, dx = \int_a^b \underbrace{\left[\frac{\partial F}{\partial y} - \frac{d}{dx} \left(\frac{\partial F}{\partial y'} \right) \right]}_{\nabla \mathcal{J}} \varphi \, dx \quad \forall \varphi \in C_0^\infty(a, b) \tag{4.4}$$

and by the fundamental lemma (3.13), we conclude the Lagrange equation is

$$\nabla \mathcal{J}(y) = \frac{\partial F}{\partial y} - \frac{d}{dx} \left(\frac{\partial F}{\partial y'} \right) = 0 \tag{4.5}$$

Equation (4.4) is also called the **weak form** of Euler-Lagrange equation. Calculating $\delta \mathcal{J}(u, \varphi)$ is relatively easy, but getting the gradient $\nabla \mathcal{J}$ can become very complicated.

Since $\delta \mathcal{J}(u; \varphi)$ is a bounded linear functional, $\nabla \mathcal{J}(u) \in \mathcal{H}$ is uniquely defined by the Riesz representation theorem. It should be noted, however, that $\delta \mathcal{J} : \mathcal{H} \to \mathcal{H}^*$ and $\nabla \mathcal{J} : \mathcal{H} \to \mathcal{H}$ are not identical; the first is a linear functional and the second an element of $\mathcal{H}$. But they are "equivalent" in the sense of (4.2), so some authors tend to use the two terms without making a careful distinction.

Depending on the context, literature, and the area we are working on, the gradient may have different names and interpretations. Sometimes the expression (4.5) is called **functional derivative**, this is precisely the left hand side of Euler-Lagrange equation (1.8). In physics textbooks it is customary to say **variational derivative** $\frac{\delta \mathcal{J}}{\delta y}$ and write

$$\delta \mathcal{J}_\varphi(y) = \int_a^b \frac{\delta \mathcal{J}}{\delta y} \varphi(x) dx$$

4.2 Local Extrema For Differentiable Functionals

Theorem 4.1 *Let $\mathcal{J}[\cdot]$ be a real valued functional with Gâteaux differential $\delta \mathcal{J}(u; \varphi)$ on the vector space $\mathcal{U}$. A necessary condition for $\mathcal{J}[\cdot]$ to have an extremum at $u_0 \in \mathcal{U}$ is that $\delta \mathcal{J}(u_0; \varphi) = 0$ for all $\varphi \in \mathcal{U}$.*

This result follows from observing that $\mathcal{J}[u + \varepsilon \varphi]$ is a function of the real variable ε with extremum in $\varepsilon = 0$. Then by (4.2) we have

$$\frac{\partial}{\partial \varepsilon} \mathcal{J}[u + \varepsilon \varphi]\bigg|_{\varepsilon=0} = 0 \qquad (4.6)$$

Example Going back to the problem (1.4) of minimum arc length

$$\mathcal{J}[y] = \int_{x_0}^{x_1} \sqrt{1 + y'^2}\, dx$$

and applying (4.6)

$$\mathcal{J}[y + \varepsilon \varphi] = \int_{x_0}^{x_1} \sqrt{1 + (y' + \varepsilon \varphi')^2}\, dx$$

$$\frac{\partial}{\partial \varepsilon} \mathcal{J}(y + \varepsilon \varphi) = \int_{x_0}^{x_1} \frac{\partial}{\partial \varepsilon} \sqrt{1 + (y' + \varepsilon \varphi')^2}\, dx$$

$$\delta \mathcal{J}(y; \varphi) = \int_{x_0}^{x_1} \frac{y'}{\sqrt{1 + y'^2}} \varphi'\, dx$$

assuming $u = \frac{y'}{\sqrt{1+y'^2}}$ and using integration by parts for $\varphi \in C_0^\infty(x_0, x_1)$

$$\delta \mathcal{J}(y; \varphi) = \int u \varphi' = -\int u' \varphi = 0 \quad \text{for all } \varphi \in C_0^\infty(x_0, x_1)$$

Now, by the fundamental lemma (3.13), $u' = 0$, thus

$$\frac{d}{dx} \frac{y'}{\sqrt{1 + y'^2}} = 0$$

and as a consequence $\frac{y'}{\sqrt{1+y'^2}} = C$. Simplifying this last expression, $y' = C$ and integrating we obtain the result we were waiting for, the equation of a line $y = ax + b$. □

4.2 Local Extrema For Differentiable Functionals

Functionals with Higher Order Derivatives

Find Euler-Lagrange equation for a functional containing a single function and some of its higher order derivatives, for example

$$\mathscr{F}[y] = \int_{x_0}^{x_1} F(x, y, y', y'')dx$$

with boundary conditions

$$y^{(k)}(x_0) = y_0^{(k)}$$

$$y^{(k)}(x_1) = y_1^{(k)} \qquad \varphi \in C_0^\infty(x_0, x_1)$$

Doing $\bar{y} = y + \varepsilon\varphi$ and $\bar{y}' = y' + \varepsilon\varphi'$, ..., $\bar{y}^{(k)} = y^{(k)} + \varepsilon\varphi^{(k)}$ with $\frac{\partial \bar{y}^k}{\partial \varepsilon} = \varphi^k$ we have

$$\mathscr{F}[y + \varepsilon\varphi] = \int F(x, \bar{y}', \bar{y}'', \bar{y}''')dx$$

$$\frac{\partial}{\partial \varepsilon}\mathscr{F}[u + \varepsilon\varphi] = \int \frac{\partial}{\partial \varepsilon} F(x, \bar{y}', \bar{y}'', \bar{y}''')dx = \int \frac{\partial F}{\partial \bar{y}}\frac{\partial \bar{y}}{\partial \varepsilon} + ... + \frac{\partial F}{\partial \bar{y}'''}\frac{\partial \bar{y}'''}{\partial \varepsilon}$$

$$\left.\frac{\partial}{\partial \varepsilon}\mathscr{F}[u + \varepsilon\varphi]\right|_{\varepsilon=0} = \int F_y\varphi + F_{y'}\varphi' + F_{y''}\varphi'' + F_{y'''}\varphi''' \qquad \forall \varphi \in C_0^\infty(x_0, x_1)$$

$$= \int \left(F_y - \frac{d}{dx}F_{y'} + \frac{d^2}{dx^2}F_{y''} - \frac{d^3}{dx^3}F_{y'''}\right)\varphi dx$$

and in general for the functional

$$\mathscr{F} = \int F(x, y, y', y'', ..., y^{(m)})$$

we obtain (**Euler-Poison**)

$$\nabla \mathscr{F} = \frac{\partial F}{\partial y} - \frac{d}{dx}\frac{\partial F}{\partial y'} + \frac{d^2}{dx^2}\frac{\partial F}{\partial y''} - ...(-1)^m \frac{d^m}{dx^m}\frac{\partial F}{\partial y^m} \qquad (4.7)$$

for the gradient of the functional.

Functionals with Several Functions

Let us consider functionals of the form

$$\mathscr{J}[y_1, y_2, ..., y_n] = \int_{t_0}^{t_1} F(t, y_1, y_2, ..., y_n, y_1', y_2', ..., y_n') dt \tag{4.8}$$

with the corresponding boundary conditions $y_k(t_0) = y_{k,0}$; $y_k(t_1) = y_{k,1}$ and

$$Y_k(t) = y_k(t) + \varepsilon_k \varphi_k(t) \qquad \varphi_k \in C_0^\infty(t_0, t_1)$$

So the integral to minimize is

$$\mathscr{J}[\varepsilon_1, ..., \varepsilon_n] = \int_{t_0}^{t_1} F(t, ..., y_k + \varepsilon_k \varphi_k, ..., y_k' + \varepsilon_k \varphi_k') dt$$

$$\frac{\partial \mathscr{J}}{\partial \varepsilon_k} = \int_{t_0}^{t_1} \frac{\partial F}{\partial \varepsilon_k} dt$$

$$\frac{\partial F}{\partial \varepsilon_k} = \frac{\partial F}{\partial Y_k} \frac{\partial Y_k}{\partial \varepsilon_k} + \frac{\partial F}{\partial Y_k'} \frac{\partial Y_k'}{\partial \varepsilon_k} = \frac{\partial F}{\partial Y_k} \varphi_k + \frac{\partial F}{\partial Y_k'} \varphi_k'$$

Replacing

$$\frac{\partial \mathscr{J}}{\partial \varepsilon_k} = \int_{t_0}^{t_1} \left(\frac{\partial F}{\partial Y_k} \varphi_k + \frac{\partial F}{\partial Y_k'} \varphi_k' \right) dt$$

and integrating by parts

$$\left. \frac{\partial \mathscr{J}}{\partial \varepsilon_k} \right|_{\varepsilon_k=0} = \int_{t_0}^{t_1} \left(\frac{\partial F}{\partial y_k} - \frac{d}{dx} \frac{\partial F}{\partial y_k'} \right) \varphi_k \qquad \varphi_k \in C_0^\infty[t_0, t_1]; k = 1, ..., n$$

So the solutions are the Euler-Lagrange equations

$$\frac{\partial F}{\partial y_k} - \frac{d}{dx} \frac{\partial F}{\partial y_k'} = 0; \qquad k = 1, ..., n \tag{4.9}$$

Functionals with Two Independent Variables

Consider the functional

$$\mathscr{J}[u] = \iint_\Omega F(x, y, u, u_x, u_y) dy\, dx$$

4.2 Local Extrema For Differentiable Functionals

proceeding in a similar way to (1.6) and applying (4.6)

$$\Phi(\varepsilon) = \mathcal{J}[u + \varepsilon\varphi] = \iint_\Omega F(x, y, u + \varepsilon\varphi, u_x + \varepsilon\varphi_x, u_y + \varepsilon\varphi_y)\,dy\,dx$$

$$\left.\frac{\partial}{\partial \varepsilon}\mathcal{J}[u + \varepsilon v]\right|_{\varepsilon=0} = \iint_\Omega \left(\varphi F_u + \varphi_x F_p + \varphi_y F_q\right) dy\,dx \qquad (4.10)$$

with

$$p = \varphi_x, q = \varphi_y$$

To identify the functional gradient, we need to rewrite this integral in the form of an inner product

$$\langle \nabla \mathcal{J}, \varphi \rangle = \iint_\Omega G(x, y)\varphi(x, y)\,dy\,dx, \qquad \text{where } G = \nabla \mathcal{J}.$$

Now we apply Green's Theorem in the form

$$\iint_\Omega \left(\frac{\partial N}{\partial x} + \frac{\partial M}{\partial y}\right) dy\,dx = \int_{\partial\Omega} -M\,dx + N\,dy$$

Making $M = f_2\varphi, N = f_1\varphi$

$$\iint_\Omega [(f_2)_y + (f_1)_x]\varphi + \iint_\Omega f_2\varphi_y + f_1\varphi_x = \int_{\varphi\Omega} -f_2\varphi\,dx + f_1\varphi\,dy = 0$$

then making $f_1 = F_p, f_2 = F_q$, we have

$$\iint \varphi_x F_p + \varphi_y F_q = -\iint \left[\frac{\partial}{\partial x}\left(\frac{\partial F}{\partial p}\right) + \frac{\partial}{\partial y}\left(\frac{\partial F}{\partial q}\right)\right]\varphi$$

and replacing in (4.10)

$$\langle \nabla \mathcal{J}, \varphi \rangle = \iint_\Omega \left[\frac{\partial F}{\partial u} - \frac{\partial}{\partial x}\left(\frac{\partial F}{\partial p}\right) - \frac{\partial}{\partial y}\left(\frac{\partial F}{\partial q}\right)\right]\varphi$$

so the gradient of the functional is

$$\nabla \mathcal{J} = \frac{\partial F}{\partial u} - \frac{\partial}{\partial x}\left(\frac{\partial F}{\partial p}\right) - \frac{\partial}{\partial y}\left(\frac{\partial F}{\partial q}\right) \qquad (4.11)$$

Example (Dirichlet Problem) Minimize the Dirichlet integral

$$\mathcal{J}[u] = \int_\Omega |\nabla u|^2 d\mathbf{x} = \iint_\Omega (u_x^2 + u_y^2) dy dx$$

$$F_u = 0; \; F_p = 2u_x; \; F_q = 2u_y;$$

then

$$\nabla \mathcal{J} = -\frac{\partial}{\partial x}(u_x) - \frac{\partial}{\partial y}(u_y) = -u_{xx} - u_{yy} = -\Delta u$$

If the integrand of the functional $\mathcal{J}$ depends on derivatives of higher order, then using successively the same line of argument as in derivation of the former cases we find that any function that makes $\mathcal{J}$. If an extremum must satisfy an analogous equation. For instance, in the case of

$$\mathcal{J}[u(x,y)] = \iint_\Omega F\left(x, y, u, u_x, u_y, u_{xx}, u_{xy}, u_{yy}\right) dy dx$$

we have the following equation

$$\nabla \mathcal{J} = F_u - \frac{\partial}{\partial x} F_{u_x} - \frac{\partial}{\partial y} F_{u_y} + \frac{\partial^2}{\partial x^2} F_{u_{xx}}$$
$$+ \frac{\partial^2}{\partial x \partial y} F_{u_{xy}} + \frac{\partial^2}{\partial y^2} F_{u_{yy}} + \cdots + (-1)^n \frac{\partial^n}{\partial y^n} F_{u_{yy\cdots y}}$$

The n-th Variation of a Functional

If we remember (from chapter1) $\Phi(\varepsilon) = \mathcal{J}[u_0 + \varepsilon\varphi], \varepsilon \in \mathbb{R}$, and (4.2), we can study the behavior of Φ around $\varepsilon = 0$ with Taylor classical theorem

$$\Phi(\varepsilon) = \Phi(0) + \sum_{k=1}^n \frac{\varepsilon^k}{k!} \Phi^k(0) + R_n$$

such that

$$\Phi^{(k)}(0) = \frac{\partial^k}{\partial \varepsilon^k} \mathcal{J}[u + \varepsilon\varphi]\bigg|_{\varepsilon=0} \quad (4.12)$$

then we define the *n*-**th variation** of the functional $\mathcal{J}$ at the point u in the direction φ by

$$\delta^n \mathcal{J}(u; \varphi) := \Phi^{(k)}(0) \quad (4.13)$$

In particular, provided $\delta \mathcal{J}$ exists, the second Gateaux variation is

$$\delta^2 \mathcal{J}(u, \varphi_1, \varphi_2) = \lim_{\varepsilon \to 0} \frac{1}{\varepsilon}[\delta \mathcal{J}(u + \varepsilon \varphi_1; \varphi_2) - \delta \mathcal{J}(u; \varphi_2)]$$

if such a limit exists.

If $\mathcal{J} : U(u_0) \subseteq \mathcal{U} \to \mathbb{R}$ is a functional on the open neighborhood $U(u_0)$ in the real normed space $\mathcal{U}$. Then the following classical necessary and sufficient conditions for local extrema holds [128].

1. *Necessary condition.* If $\mathcal{J}$ has a local minimum or a local maximum at the point u_0, then u_0 is a critical point of $\mathcal{J}$, that is,

$$\delta \mathcal{J}(u_0, h) = 0 \qquad \forall h \in \mathcal{U} \qquad (c1)$$

provided the first variation $\delta \mathcal{J}(u_0, h)$ exists for each $h \in \mathcal{U}$.

If the Gateaux derivative exists, then condition (c1) is equivalent to

$$\nabla \mathcal{J}[u_0] = 0 \qquad \text{Euler equation.}$$

2. *Sufficient condition.* The functional $\mathcal{J}$ has a local minimum at the point u_0 provided the following hold true:

- Condition (c1) is satisfied.
- The second variation $\delta \mathcal{J}(u; h)$ exits for all u in a neighborhood of u_0 and for all $h \in \mathcal{U}$. There is a constant $C > 0$ such that

$$\delta^2 \mathcal{J}(u_0; h) \geq C \|h\|^2 \qquad \text{for all } h \in \mathcal{U}$$

- For each given $\varepsilon > 0$, there exists an $\eta(\varepsilon) > 0$ such that

$$\left| \delta^2 \mathcal{J}(u; h) - \delta^2 \mathcal{J}(u_0; h) \right| \leq \varepsilon \|h\|^2$$

for all $u, h \in \mathcal{U}$ with $\|u - u_0\| \leq \eta(\varepsilon)$.

4.3 Geometry of Surfaces

Surface geometry [33, 71, 85] plays a crucial role in computer vision, influencing various aspects of image understanding, scene analysis, and 3D reconstruction. Understanding the geometry of surfaces enables computer vision systems to extract meaningful information from images, infer spatial relationships, and make accurate interpretations of the scene. So the first thing we need is a surface representation.

Surfaces can be described using different mathematical representations, each offering unique advantages and applications.

In the *explicit form* $z = f(x, y)$, a surface is represented explicitly as a function of x and y and where z represents the height or elevation of the surface at each point (x, y) in the plane. This representation is particularly useful when the surface can be directly expressed in terms of x and y, such as in the case of terrain maps, elevation models, or mathematical functions representing surfaces. For example, the equation $z = f(x, y)$ could represent the surface of a hill, where x and y are the coordinates on the ground plane, and z represents the elevation above sea level.

In the *implicit form*, a surface is represented by an equation of the form $f(x, y, z) = c$, where $f(x, y, z)$ is a scalar function that describes the surface, and c is a constant. This equation defines a level set of the function $f(x, y, z)$ at the value c, which corresponds to the surface. The implicit form is particularly useful when describing surfaces that are not easily parameterized or when the relationship between the surface and its parameters is complex. For example, the equation of a sphere centered at the origin with radius r can be represented in the implicit form as $x^2 + y^2 + z^2 = r^2$.

In *parametric form*, a surface is described by a set of parametric equations that define the coordinates of points on the surface as functions of two parameters u and v. These equations are represented as a vector-valued function

$$\mathbf{x}(u, v) = (x(u, v), y(u, v), z(u, v))$$

The parameters u and v vary over a certain domain, often corresponding to a region in the plane. By varying u and v, different points on the surface are generated, allowing for a more flexible representation of the surface (Fig. 4.1). The parametric form is widely used in differential geometry and computer graphics for its flexibility and ease of manipulation. For example, the parametric equations of a sphere centered at the origin with radius $r = 1$ can be given as

$$\mathbf{x}(u, v) = (\cos u \sin v, \sin u \sin v, \cos v)$$

with $(u, v) \in [0, 2\pi] \times [0, \pi]$.

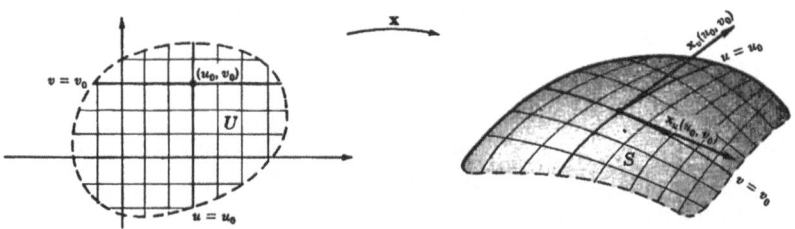

Fig. 4.1 Curvilinear coordinates

4.3 Geometry of Surfaces

Once we have properly defined a tangent plane on a point P of the surface Σ, can be developed a surface theory. We do not intend to reproduce the theory of surfaces here, but in general terms the idea is the following.

If C is a curve $w(t) = (u(t), v(t))$ in the parameter plane, then $\mathbf{r}(t) = \mathbf{x}(u(t), v(t))$ is a curve on Σ. We want to find the tangent vector $\mathbf{r}'(t)$, then applying chain rule in the components of $\mathbf{r}(t)$

$$\mathbf{r}(t) = \mathbf{x}(u(t), v(t)) = (x(u(t), v(t)), y(u(t), v(t)), z(u(t), v(t)))$$

$$\mathbf{r}'(t) = \frac{d}{dt}(x(u(t), v(t)), ..., ...) = (x_u \frac{du}{dt} + x_v \frac{dv}{dt}, ..., ...)$$

$$= \frac{du}{dt}(x_u, y_u, z_u) + \frac{dv}{dt}(x_v, y_v, z_v)$$

$$= \frac{du}{dt}\mathbf{x}_u + \frac{dv}{dt}\mathbf{x}_v$$

So we can represent the tangent vector as

$$d\mathbf{x} = \mathbf{x}_u du + \mathbf{x}_v dv \tag{4.14}$$

If $\mathbf{x} = \mathbf{x}(u(t), v(t))$ is a regular arc on a patch $\mathbf{x} = \mathbf{x}(u, v)$ its length is the integral

$$s = \int_a^b \left|\frac{d\mathbf{x}}{dt}\right| dt = \int_a^b \langle d\mathbf{x}, d\mathbf{x}\rangle^{1/2} dt$$

$$\langle d\mathbf{x}, d\mathbf{x}\rangle = \left\langle \mathbf{x}_u \frac{du}{dt} + \mathbf{x}_v \frac{dv}{dt}, \mathbf{x}_u \frac{du}{dt} + \mathbf{x}_v \frac{dv}{dt}\right\rangle = E\left(\frac{du}{dt}\right)^2 + 2F \frac{du}{dt}\frac{dv}{dt} + G\left(\frac{dv}{dt}\right)^2$$

with

$$E = \mathbf{x}_u \cdot \mathbf{x}_u, \qquad F = \mathbf{x}_u \cdot \mathbf{x}_v, \qquad G = \mathbf{x}_v \cdot \mathbf{x}_v$$

$$s = \int_a^b (E du^2 + 2F du dv + G dv^2)^{1/2} dt$$

Thus the **first fundamental form** I is the quadratic form defined on vectors (du, dv) in the uv plane by

$$I(du, dv) = E du^2 + 2F du dv + G dv^2 \tag{4.15}$$

The first fundamental form of a surface is a mathematical construct that describes the intrinsic geometry of the surface. It provides essential information about the lengths of curves, angles between curves, and the metric properties of the surface.

Minimal Surfaces

The first systematic study of minimal surfaces is attributed to Leonhard Euler in the eighteenth century. Euler investigated surfaces of minimal area subject to various boundary conditions, laying the groundwork for the calculus of variations. He introduced the concept of "geodesic curvature" and derived the famous Euler-Lagrange equation for minimal surfaces [50].

In its simplest form [109], Euler proposed to find the surface of smallest area bounded by a given closed curve in space. If we assume that this curve projects down to a closed curve C surrounding a region Ω in the xy plane, and also that the surface is expressible in the form $u = u(x, y)$, then the problem is to minimize the surface area integral

$$\mathcal{J}[u] = \iint_\Omega \sqrt{1 + u_x^2 + u_y^2}\, dy dx, \qquad p = u_x, \; q = u_y$$

$$\frac{\partial F}{\partial u} = 0; \quad \frac{\partial F}{\partial p} = \frac{p}{\sqrt{1 + p^2 + q^2}}; \quad \frac{\partial F}{\partial q} = \frac{q}{\sqrt{1 + p^2 + q^2}}$$

$$-\frac{\partial}{\partial x} \frac{u_x}{\sqrt{1 + u_x^2 + u_y^2}} - \frac{\partial}{\partial y} \frac{u_y}{\sqrt{1 + u_x^2 + u_y^2}}$$
$$= \frac{-(1 + u_y^2)u_{xx} + 2u_x u_y u_{xy} - (1 + u_x^2)u_{yy}}{(1 + u_x^2 + u_y^2)^{3/2}} = 0$$

So the Euler's equation has the form

$$(1 + u_y^2)u_{xx} - 2u_x u_y u_{xy} + (1 + u_x^2)u_{yy} = 0 \qquad (4.16)$$

Historically, one of the first results that drew attention to minimal surfaces is the relationship of this formula with other ideas of surface geometry; for example the mean curvature H

$$H = \frac{EN - 2FM + GL}{2(EG - F^2)}$$

where E, F, G are the coefficients of the first fundamental form of the surface and L, M, N, the coefficients of its second fundamental form (see [71]). Calculating, we can obtain

$$E = 1 + u_x^2, \qquad F = u_x u_y, \qquad G = 1 + u_y^2,$$

$$L = \frac{u_{xx}}{\sqrt{1 + u_x^2 + u_y^2}}, \qquad M = \frac{u_{xy}}{\sqrt{1 + u_x^2 + u_y^2}}, \qquad N = \frac{u_{yy}}{\sqrt{1 + u_x^2 + u_y^2}}$$

4.3 Geometry of Surfaces

Fig. 4.2 Catenoid
$(2\cos u \cosh(v/2), 2\sin u \cosh(v/2), v)(u, v) \in [0, 2\pi] \times [-3, 3]$

thus

$$H = \frac{(1+u_y^2)u_{xx} - 2u_x u_y u_{xy} + (1+u_x^2)u_{yy}}{(1+u_x^2+u_y^2)^{3/2}}$$

and Eq. (4.16) implies that the mean curvature H of the solution surface is zero. So a **minimal surface** is defined as a surface with zero mean curvature everywhere (Fig. 4.2).

If the integrand of the functional $\mathscr{J}$ depends on derivatives of higher order, the same line of argument as in (4.10) can be used. For example, for the functional

$$\mathscr{J}[u] = \iint_\Omega F\left(x, y, u, \frac{\partial u}{\partial x}, \frac{\partial u}{\partial y}, \frac{\partial^2 u}{\partial x^2}, \frac{\partial^2 u}{\partial y \partial x}, \frac{\partial^2 u}{\partial y^2}\right)$$

$$\nabla \mathscr{J} = F_u - \frac{\partial}{\partial x}F_{u_x} - \frac{\partial}{\partial y}F_{u_y} + \frac{\partial^2}{\partial x^2}F_{u_{xx}} + \frac{\partial^2}{\partial y \partial x}F_{u_{xy}} + \frac{\partial^2}{\partial y^2}F_{u_{yy}} \quad (4.17)$$

Example The thin plate functional

$$\mathscr{J}[u] = \int_\Omega (u_{xx}^2 + 2u_{xy}^2 + u_{yy}^2)dydx \quad (4.18)$$

have been widely studied in computer vision problems, especially in surface reconstruction. Applying the functional gradient (4.17), we obtain

$$\nabla \mathscr{J} = \frac{\partial^2}{\partial x^2}(2u_{xx}) + \frac{\partial^2}{\partial x \partial y}(4u_{xy}) + \frac{\partial^2}{\partial y^2}(2u_{yy})$$

$$= 2\left(\frac{\partial^4 u}{\partial x^4} + \frac{\partial^4 u}{\partial x^2 \partial y^2} + \frac{\partial^4 u}{\partial y^4}\right)$$

$$= 2\Delta(\Delta u)$$

where $\Delta u = u_{xx} + u_{yy}$ is the Laplacian operator and

$$\Delta^2 u := \Delta(\Delta u) = \frac{\partial^4 u}{\partial x^4} + \frac{\partial^4 u}{\partial x^2 \partial y^2} + \frac{\partial^4 u}{\partial y^4} \tag{4.19}$$

is the Biharmonic operator

□

The Gradient Algorithm for Functionals

Gradient descent is a global optimization method that can be formulated in a Banach space setting. Commonly these optimization methods are iterative algorithms for finding (global or local) solutions of minimization problems. For minimizing $\mathcal{J}$: $U \subset \mathcal{U} \to \mathbb{R}$ continuously differentiable in the Banach space $\mathcal{U}$ the method generates a sequence of iterates $\{u_k\} \subset U$ to solve the unconstrained optimization problem

$$\min_{u \in \mathcal{U}} \mathcal{J}(u)$$

The first-order optimality condition for a local minimum u is well-known: $\nabla \mathcal{J} = 0$. The idea of descent methods is to find, at the current k-th iterate $u_k \in \mathcal{U}$, a direction w_k such that $\Phi_k(\tau) := \mathcal{J}(u_k + \tau w_k)$ is decreasing at $\tau = 0$, then

$$\Phi'_k(0) = \langle \nabla f(u_k), w_k \rangle_{\mathcal{U}^* \times \mathcal{U}} < 0$$

Algorithm 1 Gradient descent for functionals

1: Input: Choose an initial point $u_0 \in \mathcal{U}$
2: **for** $k = 1, 2, \ldots$ **do**
3:     If $\nabla \mathcal{J}(u_k) = 0$, STOP
4:     Choose a descent direction $w_k \in \mathcal{U}$ such that $\langle \nabla f(u_k), w_k \rangle < 0$ for example $w_k = -\nabla f(u_k)$
5:     $u_{k+1} = u_k - \tau \nabla f(u_k)$
6: **end for**
7: Output:x_{k+1}

Chapter 5
Inverse Problems and Variational Regularization

Inverse problems arise in many areas of image processing, where the goal is to recover unknown quantities from indirect, incomplete, or noisy measurements. These problems are typically ill-posed, meaning that small changes in the data can lead to large changes in the solution, or that solutions may not exist or be unique. To overcome these challenges, variational regularization methods have become essential. In this chapter, we introduce the foundational concepts of inverse problems, discussing their mathematical formulation, including examples from image deblurring and denoising. We then present variational regularization as a powerful framework to stabilize these problems by incorporating prior information about the solution, such as smoothness or sparsity. The chapter explores classical regularization techniques like Tikhonov regularization and Total Variation (TV), along with more advanced models, and examines their mathematical properties (like convexity) and practical applications. We also discuss the role of regularization parameters, the construction of energy functionals by their physical properties and the corresponding Euler-Lagrange equations, which serve as the basis for designing efficient numerical algorithms. Through a combination of theory and examples, this chapter provides a comprehensive overview of how variational regularization methods address inverse problems in image processing.

5.1 Linear Inverse Problems

To put it simply, an inverse problem consists in determining the causes from the measurable effects of a phenomenon. The greatest difficulty is that we often cannot directly measure the cause we are looking for. It is natural that if we talk about an inverse problem we should be able to talk about a forward problem. In fact, the classical problems of physical mathematics are direct forward problems, since their purpose is to create a mathematical model that predicts the result of an experiment, assuming that all experimental conditions are known.

In the case of image processing, there is a detector that registers values of $u(t)$; however, what the detector produces is a more or less faithful f version of the real scene defined by

$$f(s) = \frac{1}{\sqrt{2\pi}\sigma} \int e^{-(s-t)^2/(2\sigma^2)} u(t) dt$$

the direct problem is to model the behavior of the detector and produce this formula. The result of the experiment is the registration of the image f from the true image u. The inverse problem is to estimate the true image u from the measured data f.

The most important aspect of direct problems is **stability**, which can be defined by the concept of continuity of a function, that is, if two points in the domain are close, their images are too. In other words, small errors in the measurement of u do not significantly affect the calculation of f. On the contrary, the characteristic of an inverse problem is its **instability**: the quantity u that we intend to measure does not depend continuously on the measured data f. This is mainly due to noise present in the data, yet it is possible to reconstruct the unknown solution. The art of carrying out this reconstruction is called **regularization**.

A **linear inverse problem** can be represented as

$$\mathcal{K}u = f \tag{5.1}$$

with $\mathcal{K} : \mathcal{U} \to \mathcal{V}$ a linear operator (called the **forward operator**), f is a given (data) vector, and the goal is to determine u. The main problem is that u does not depend continuously on f, which means that if $\mathcal{K}^{-1}$ exists it is a discontinuous operator.

5.2 Image Deblurring as an Inverse Problem

The image of a scene or an object captured by a camera and other optical instruments is always imperfect. If the object is represented by a function $u : \mathbb{R}^2 \to \mathbb{R}$ that represents intensity or grayscales and the image produced by the device is the function $f : \mathbb{R}^2 \to \mathbb{R}$.

Then in this process $\mathcal{K}$ is an operator that models the construction of the image [104]. If $\mathcal{K}$ were the perfect or ideal operator then $\mathcal{K} = I$ is the identity operator. Therefore the perfect model can be represented in terms of a Dirac delta distribution $\langle \delta_{(\mathbf{x})}, u \rangle = u(\mathbf{x})$ or

$$f(\mathbf{x}) = \int_{\mathbb{R}^2} \delta(\mathbf{x} - \mathbf{t}) u(\mathbf{t}) \, d\mathbf{t}$$

5.2 Image Deblurring as an Inverse Problem

However, regardless of the physical device used, the image obtained f^δ is a blurred image of the object u. When the blur is linear and shift invariant is reduced (represented) to convolution operators, and deblurring is therefore called deconvolution. Sometimes f^δ is represented as

$$f^\delta(\mathbf{x}) = \int_{\mathbb{R}^2} k(\mathbf{x} - \mathbf{t}) u(\mathbf{t})\, d\mathbf{t} \tag{5.2}$$

where $k(\cdot)$ is the *point spread function* of the device and is some approximation of the Dirac delta δ.

Given the blurred image f^δ the inverse problem for **deblurring** is to find u by solving (5.2). This type of problem in which a direct calculation is not carried out, but rather requires the inversion of a process, frequently occurs in the world of images. In other words, now we are interested in finding a suitable input u that produces the given output f^δ.

In other words, given $\mathcal{K}u = f$, the objective is to determine u given f or its perturbed version f^δ

$$\|f - f^\delta\| \leq \delta,$$

where δ can be thought of as the noise strength and is a small positive quantity. If the nature of $\mathcal{K}$ is such that some information is lost in the forward process of formation of f from u, then the inverse problem of determining u from f becomes an **ill-posed problem**. Depending on whether $\mathcal{K}$ is a linear or non-linear operator, the ill-posed problem is classified as linear or non-linear, respectively.

In general, if η^δ represents additive noise in the data f^δ and k is the point spread function, f^δ and u can be related by the degradation model

$$f^\delta(\mathbf{x}) = \int_\Omega k(\mathbf{x}, \mathbf{y}) u(\mathbf{y}) d\mathbf{y} + \eta^\delta(\mathbf{x}); \qquad \mathbf{x} \in \Omega \tag{5.3}$$

In our applications, k is smooth, and, hence, the integral operator is compact. Moreover, k does not depend on $\mathbf{x}$ and $\mathbf{y}$ individually but only on $\mathbf{x} - \mathbf{y}$. Thus, with a slight abuse of notation, we have $k(\mathbf{x}, \mathbf{y}) = k(\mathbf{x} - \mathbf{y})$. Then (5.3) can be expressed as

$$f^\delta = k * u + \eta^\delta \tag{5.4}$$

Our task is to recover u given k and the observed image f^δ. A Straightforward solution of the linear model

$$f^\delta = k * u \tag{5.5}$$

for u does not provide a meaningful approximation of the desired noise- and blur-free image. The reason for this is that (5.5) ignores the noise η^δ in the right-hand side of (5.4) and that the inverse of the integral operator in (5.5) is unbounded if it exists. The latter follows from the compactness of the integral operator. The task of solving (5.5) therefore is an ill-posed problem [38].

5.3 Regularization Methods

Regularization methods aim to find stable approximate solutions of inverse and ill-conditioned problems. Inverse problems arise not only in image processing, but also in a wide variety of applications. There is a great need for stable solutions to these problems as well as the convergence of these solutions. All these maps can be characterized by operator equations (5.1).

The goal of regularization is to construct an operator $\mathcal{R} : \mathcal{V} \to \mathcal{U}$ such that $u = \mathcal{R}f$ is an approximate solution of $\mathcal{K}u = f$. In the most important applications is not possible to find a single operator $\mathcal{R}$ that produce an acceptable answer for all $f \in \mathcal{V}$. In these cases we can construct a family $\{\mathcal{R}_\lambda : \lambda > 0\}$ such that $u = \mathcal{R}_\lambda f$ is an acceptable solution for some $\lambda > 0$.

In general the range of $\mathcal{K}$ is not closed, so these equations are ill conditioned in the sense that very small perturbations in f can cause very large changes in the solution u. To achieve stable approximate solutions, regularization methods must be applied, which attenuate the discontinuity of the inverse function $\mathcal{K}^{-1}$.

Inverse problems are inherently difficult to solve [7], owing to the fact that there is a loss of information as in the case when $\mathcal{K}$ is equivalent to low pass filtering where the high frequency information is lost. Inverse problems are typically ill-posed. A problem is **well-posed** in the Hadamard sense [105] if the following three conditions are satisfied:

1. A solution exists.
2. The solution is unique.
3. The solution is stable under perturbation of data, i.e., the solution depends continuously on data, or an infinitesimal change in the data should produce only a similar change in the solution.

For the inverse problem given by $\mathcal{K}u = f$, the first condition is equivalent to the forward operator $\mathcal{K}$ being surjective. The condition of unique solution maps to $\mathcal{K}$ being injective. First and second conditions together imply that the inverse exists and the third condition requires the inverse to be a continuous function. Of the three conditions, violation of the third leads to numerical instabilities. There are no mathematical methods to convert an inherently unstable problem to a stable one. The best that can be done is to solve an approximation to the problem by using regularizers. Regularization methods recover partial information about the solution as stably as possible [121]. Violation of first and second conditions does not pose serious problems as the violation of the third [38].

5.3 Regularization Methods

Variational Approach for Constructing Regularizing Algorithm

For the stable approximation of a solution of the operator equation $\mathcal{K}u = f$ **variational regularization** methods propose to minimize the functional

$$\mathcal{T}[u] := \underbrace{\rho(\mathcal{K}u, f^\delta)}_{\text{data reproduction}} + \lambda \underbrace{\mathcal{R}(u)}_{\text{generalization}} \tag{5.6}$$

where $\rho(\cdot, \cdot)$ is the *discrepancy functional* or *data fidelity term*, measuring the error between $\mathcal{K}u$ and f^δ, $\lambda > 0$, and the *penalty term* or *regularization functional* $\mathcal{R}(\cdot)$ is a non-negative functional. The number λ is called the *regularization parameter*. In most regularization applications for the solution of inverse problems ρ is

$$\rho(\mathcal{K}u, f^\delta) = \int_\Omega (\mathcal{K}u - f^\delta)^2 \, d\mathbf{x}$$

For the case of deblurring problem a meaningful approximation of u can be determined by minimizing the functional

$$\mathcal{T}[u] = \int_\Omega (h * u - f^\delta)^2 dx + \lambda \mathcal{R}(u) \tag{5.7}$$

In variational calculus and inverse problems, regularization functionals are essential to impose additional constraints or incorporate prior information to stabilize the solution and make it well-posed. Here are some of the most common regularization functionals [49]

1. $\mathcal{L}^1$ **Regularization**, often used in sparse reconstruction problems, encourages sparsity in the solution. The regularization functional is given by

   $$\mathcal{R}(u) = |u|_{\mathcal{L}^1} = \int_\Omega |u| d\mathbf{x}$$

2. **Tikhonov ($\mathcal{L}^2$) Regularization**
   The Tikhonov regularization is one of the most widely used regularization techniques. It involves adding a term to the functional that penalizes large values of the solution. The regularization functional is given in simple form by:

   $$\mathcal{R}(u) = \|u\|^2_{\mathcal{L}^2} = \int_\Omega u^2 d\mathbf{x}$$

3. **Total Variation** (TV) Regularization Total Variation regularization is particularly useful in image processing for preserving edges while reducing noise. It is defined as

$$\mathcal{R}(u) = \int_\Omega |\nabla u| d\mathbf{x} \tag{5.8}$$

where $|\nabla u|$ denotes the gradient magnitude of u. This functional encourages piecewise smooth solutions.

4. H^1 or **Sobolev Regularization** is used to penalize the gradient of the solution, promoting smoothness. It is defined as

$$\mathcal{R}(u) = |\nabla u|^2_{\mathcal{L}^2} = \int_\Omega |\nabla u|^2 \tag{5.9}$$

5. **Higher order functionals**
Variational models like the ROF model for denoising, often use first-order derivatives to penalize rapid changes in intensity. While effective, these models can produce piecewise constant solutions that are not always suitable for images with smooth transitions or complex textures. Higher order variational methods address this by incorporating higher-order derivatives [73] (e.g., second-order or beyond), which better capture the smoothness and curvature of the image. A general higher-order regularization functional can be formulated as

$$\mathcal{R}(u) = \int_\Omega F(\mathbf{x}, u(\mathbf{x}), \nabla u(\mathbf{x}), \nabla^2 u(\mathbf{x}), \cdots) d\mathbf{x}$$

5.4 Functionals with Physical Interpretation

Choosing the appropriate regularization functional for inverse problems involves considering several factors related to the characteristics of the problem and the desired properties of the solution. One of the most frequent problems is that different regularization functionals may encode different prior beliefs about the underlying image. For example, Total Variation regularization is effective for preserving edges and piecewise smoothness, while Tikhonov regularization favors smooth solutions.

Regularization functionals may handle noise differently. TV regularization is robust to certain types of noise, while other functionals may be more sensitive to noise amplification. Some functionals may lead to computationally intensive optimization problems, especially for large-scale images. So it is necessary a balance between incorporating these different facts. Experimentation and testing on representative datasets may be necessary to determine the most suitable regularization functional for a particular problem.

Some of the most frequent models for selecting penalty functionals come from elasticity theory [44, 116]. The methods of relating geometry to structural analogies involves drawing parallels between geometric objects and physical or mechanical systems. This approach helps to intuitively understand geometric concepts by visualizing them in terms of familiar physical phenomena. In elasticity theory, plates

5.4 Functionals with Physical Interpretation

and membranes represent different idealizations of thin structural elements. The main difference between plates and membranes lies in the way they deform under loads and the associated assumptions made in their modeling.

By a **membrane** [72] is meant a thin film offering no resistance to bending, but acting only in tension. A body having the middle surface in the form of a plane and whose thickness is sufficiently small compared with its other two dimensions is called a **thin plate**.

Membranes are assumed to be two-dimensional structures with negligible thickness compared to their other dimensions. Membranes deform primarily by stretching and shearing, with no bending. Bending is considered negligible.

Plates are also two-dimensional structures, but their thickness is not necessarily negligible. Plates deform by bending as well as stretching and shearing. Bending is a significant mode of deformation. Plates find wide application in engineering; as typical examples we may mention concrete and reinforced concrete plates used in structures, for ship hulls. A plane dividing the thickness of the plate in half is called its middle plane.

Both membranes and plates are part of the broader field of shell structures in elasticity theory, which includes thin-walled structures that are modeled based on the dominant modes of deformation. The choice between modeling a structure as a membrane or a plate depends on the expected deformation behavior and the relevance of bending effects in the particular application [95, 111].

The membrane model represents a surface as if it were a thin, flexible membrane. The displacement field of the membrane is denoted by $u(x, y)$, where x and y are spatial coordinates. The deformation of the membrane is described by the membrane's displacement field $u(x, y)$, and the equilibrium state is determined by minimizing the membrane's potential energy functional, which may have terms related to stretching or bending energy, and it can be expressed as an integral over the surface area

$$\mathcal{R}[u] = \int_\Omega F(u, u_x, u_y, u_{xx}, u_{yy}, ...) d\mathbf{x}$$

The potential energy functional is minimized to find the equilibrium state.

The theories of **plates and membranes** [96] have great complexity and their application is really an oversimplification of more general models, but even so, their success has been remarkable, which allows us to create metaphors and physical interpretations that can be generalized to other phenomena and applications.

The energy of a thin plate is given by

$$\mathcal{J} = \int_\Omega a(\kappa_1^2 + \kappa_2^2) + 2(1 - b)\kappa_1 \kappa_2 d\omega \tag{5.10}$$

where κ_1, κ_2 are the principal curvatures of the surface; a, b are constants describing properties of the material of the thin plate like resistance to bending and sheering.

Applying the properties of Gaussian $K = \kappa_1 \kappa_2$ and mean $H = \frac{1}{2}(\kappa_1 + \kappa_2)$ curvatures; the functional (5.10) is written as

$$\mathcal{J} = \int_\Omega 4aH^2(u) + 2(1 - a - b)K(u) dS$$

Although this functional gives us an exact version of the energy, it is very complex and difficult to manage [27, 53].

Often used is the so called thin **plate energy functional**. There is an exact and a simplified version of the thin plate energy. For a parametric surface $\mathbf{x} : \Omega \to \mathbb{R}^3$ these energies can be written as

$$\mathcal{J}_{exact}(\mathbf{x}) = \int_\Omega \kappa_1^2 + \kappa_2^2 \, dS$$

$$\mathcal{J}_{simple}(\mathbf{x}) = \int_\Omega \mathbf{x}_{uu}^2 + 2\mathbf{x}_{uv}^2 + \mathbf{x}_{vv}^2 \, du dv$$

The application of membrane and plate models involves the formulation of potential energy functionals that describe the behavior of surfaces. These functionals incorporate terms related to stretching and bending energy, and their minimization yields the equilibrium state of the surface. These models are fundamental in surface modeling for capturing the physical behavior of thin, flexible structures and are widely utilized in fields ranging from engineering to computer graphics.

Deblurring with Sobolev Energy

A well known choice for $\mathcal{R}(u)$ is the Sobolev functional (5.9), then for deblurring we have to minimize the functional

$$\mathcal{J} = \int_\Omega (h*u - f^\delta)^2 d\mathbf{x} + \lambda \int_\Omega |\nabla u|^2 d\mathbf{x} \quad (5.11)$$

$$= \int_\Omega (h*u - f^\delta)^2 d\mathbf{x} + \lambda \int_\Omega (u_x^2 + u_y^2) d\mathbf{x}$$

This is a quadratic functional, so it is possible to translate the problem to Fourier domain. Applying Plancherel's identity (3.32) and properties of Fourier transform

$$\mathcal{J}[\widehat{u}] = \int_\Omega (\widehat{h*u} - \widehat{f}^\delta)^2 d\xi + \lambda \int_\Omega (\widehat{u}_x^2 + \widehat{u}_y^2) d\xi$$

$$= \int_\Omega (\widehat{u} \cdot \widehat{h} - \widehat{f}^\delta)^2 d\xi + \lambda \int_\Omega \widehat{u}^2(\xi_1^2 + \xi_2^2)$$

$$= \int_\Omega (\widehat{u} \cdot \widehat{h} - \widehat{f}^\delta)^2 d\xi + \lambda \int_\Omega \widehat{u}^2 \|\xi\|^2$$

5.4 Functionals with Physical Interpretation

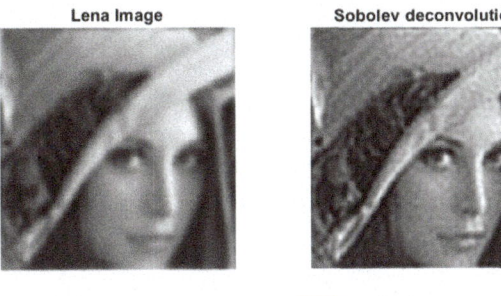

Fig. 5.1 Sobolev deconvolution (5.12) is applied to a noisy image(left)

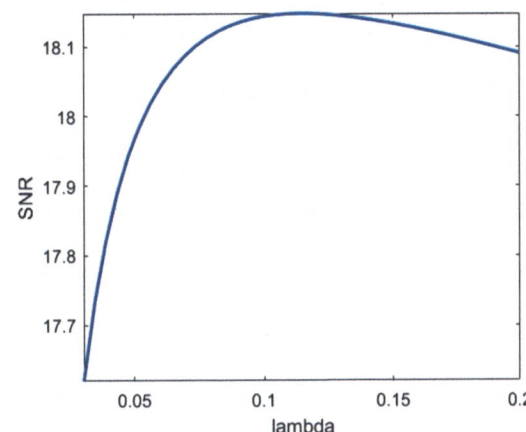

Fig. 5.2 Signal to noise ratio (SNR) depending on the regularization parameter λ in Fig. 5.1

So applying Euler-Lagrange equation (4.11) on $\mathscr{T}[\widehat{u}]$

$$(\widehat{u} \cdot \widehat{h} - \widehat{f}^\delta)\widehat{h} + \lambda \widehat{u} \|\boldsymbol{\xi}\|^2 = 0$$

and

$$\widehat{u} = \frac{\widehat{f}^\delta \widehat{h}}{\widehat{h}^2 + \lambda \|\boldsymbol{\xi}\|^2} \tag{5.12}$$

Example (Sobolev Deconvolution) Now we illustrate an application of the regularization model (5.11) translated to the Fourier domain (5.12) (Figs. 5.1 and 5.2).

Listing 5.1 Sobolev deconvolution

```
S = (X.^2 + Y.^2)*(2/n)^2;
lambda = 0.2;
fSob = real( ifft2( yF .* hF ./ ( abs(hF).^2 + lambda*S) ) );
imshow(y);
imshow(fSob)
```

5.5 Total Variation

Regularization can be seen as a tool to control the oscillations in the solution to an inverse problem. One way is to apply the total variation as penalty functional. However the total variation (2.24) as appears in mathematical analysis books $TV(f) = \sum_k |f_{k+1} - f_k|$; can be modified as

$$TV(f) = \sum_k |f_{k+1} - f_k| = \sum_k \left|\frac{f_{k+1} - f_k}{\Delta x}\right| \Delta x$$

$$\approx \int |f'| dx$$

So a natural generalization to more dimensions is the following

Definition 5.1 Let $u \in C^1(\Omega, \mathbb{R}^N)$ be a differentiable function, then

$$TV[u] = \int_\Omega \|\nabla u\|_2 d\mathbf{x}$$

is called the Total Variation of u on domain Ω. ∇u is the gradient of u.

In discrete form, for a digital image represented by a matrix u (assuming grayscale for simplicity), the total variation can be approximated as:

$$TV(u) = \sum_{i,j} \|(\nabla u)_{i,j}\|$$

where $(\nabla u)_{i,j}$ is the gradient of u at pixel (i, j).

TV regularization has a rich history in image processing, dating back to the early 1990s [17, 19]. The seminal work [102] by Rudin, Osher and Fatemi (ROF) is often considered the starting point. This work introduced TV regularization as a method for image denoising. The approach was to minimize the total variation of the image subject to fidelity constraints. ROF formulated the following minimization problem:

$$\min_u \int_\Omega \|\nabla u\| \quad s.t. \quad \|u - f\|_2^2 \leq \sigma^2 \qquad (5.13)$$

Rather than solving the constrained minimization problem (5.13), ROF and subsequent researchers also formulated an unconstrained minimization problem which uses the TV term as regularization functional

$$\min_u \left\{\int_\Omega (u - f^\delta)^2 dx + \lambda \int_\Omega \|\nabla u\|\right\} \qquad (5.14)$$

Sobolev norm (5.9) perform a denoising but also tends to blur the edges, thus producing a poor results on cartoon images. The TV prior is able to better

5.5 Total Variation

reconstruct sharp edges. With respect to the Sobolev energy, it simply corresponding to measuring the $\mathcal{L}^1$ norm instead of the $\mathcal{L}^2$ norm, thus dropping the square in the functional. Unfortunately, the TV functional is not a smooth function of the image u It thus requires the use of advanced convex optimization method to be minimized for regularization. An alternative is to replace the absolute value by a smooth absolute value.

In order to deal with the non-differentiability of the TV-norm suggested a slight perturbation of the TV-norm:

$$TV_\varepsilon = \int_\Omega \sqrt{\|\nabla u\|^2 + \varepsilon} \, d\mathbf{x} \tag{5.15}$$

where $\varepsilon > 0$ is a small positive number. The choice of ε is a trade-off between the numerical stability of the method and the accuracy of the result. Total variation is the most well known example of non-quadratic regularization [12]. A little change in the formula makes it very different in its geometric and physical interpretations.

Properties of the Total Variation Functional

1. **Convexity**. The Total Variational functional is convex. By using the triangle inequality

$$\begin{aligned} TV(\alpha u_1 + (1-\alpha)u_2) &= \int_\Omega \|\nabla(\alpha u_1 + (1-\alpha)u_2)\| d\mathbf{x} \\ &\leq \int_\Omega (|\alpha \nabla u_1| + |(1-\alpha)\nabla u_2|) \\ &\leq \alpha TV(u) + (1-\alpha)TV(u) \end{aligned}$$

2. **Variational Derivative of the ROF Model**. The following is a proof adapted from [24]:

   For $N = 2$; $\|\nabla u\| = (u_x^2 + u_y^2)^{1/2}$
   $f(\varepsilon) := \|\nabla(u + \varepsilon\varphi)\| = ((u_x + \varepsilon\varphi_x)^2 + (u_y + \varepsilon\varphi_y)^2)^{1/2}$
   Now let $f(\xi)$ be the univariate function $f(\xi) = \left[(x + c_1\xi)^2 + (y + c_2\xi)^2\right]^p$
   and $f'(\xi) = p\,[*]^{p-1}\,[2(x + c_1\xi)c_1 + (y + c_2\xi)c_2]$. By applying Taylor series for f around $\xi = 0$

$$f(\xi) = f(0) + f'(0)\xi + O(\xi^2)$$
$$f(0) = (x^2 + y^2)^p$$
$$f'(0) = p(x^2 + y^2)^{p-1}(2xc_1 + 2yc_2)$$

we have

$$f(\xi) = (x^2 + y^2)^p + p\frac{2xc_1 + 2yc_2}{(x^2 + y^2)^{1-p}}\xi + O(\xi^2)$$

Using this result with $p = 1/2$

$$f(\varepsilon) = (u_x^2 + u_y^2)^p + p\frac{2u_x\varphi_x + 2u_y\varphi_y}{(u_x^2 + u_y^2)^{1-p}}\varepsilon + O(\varepsilon^2)$$

$$= \|\nabla u\| + \frac{1}{\|\nabla u\|}(\nabla u \cdot \nabla\varphi)\varepsilon + O(\varepsilon^2)$$

$$\|\nabla(u + \varepsilon\varphi)\| - \|\nabla u\| = \frac{\nabla u \cdot \nabla\varphi}{\|\nabla u\|}\varepsilon + O(\varepsilon^2)$$

Now, we apply this result to calculate the gradient of the ROF (Rudin-Osher-Fatemi) functional on $\Omega \subset \mathbb{R}^2$

$$\mathcal{J}[u] = \int_\Omega (u - f^\delta)^2 d\mathbf{x} + \lambda \int_\Omega \|\nabla u\| d\mathbf{x} \tag{5.16}$$

In order to obtain the first variation of the functional we have

$$\mathcal{J}[u + \varepsilon\varphi] = \int_\Omega \left[(u - f^\delta)^2 + 2(u - f^\delta)\varepsilon\varphi + \varepsilon^2\varphi^2\right] d\mathbf{x}$$

$$+ \lambda \int_\Omega \|\nabla(u + \varepsilon\varphi)\| d\mathbf{x}$$

$$\mathcal{J}[u + \varepsilon\varphi] - \mathcal{J}[u] = \int_\Omega \left[2(u - f^\delta)\varepsilon\varphi + \varepsilon^2\varphi^2\right] d\mathbf{x}$$

$$+ \lambda \int_\Omega \underbrace{[\|\nabla(u + \varepsilon\varphi)\| - \|\nabla u\|]}_{\frac{\nabla u \cdot \nabla\varphi}{\|\nabla u\|}\varepsilon + O(\varepsilon^2)} d\mathbf{x}$$

then

$$\frac{1}{\varepsilon}[\mathcal{J}[u + \varepsilon\varphi] - \mathcal{J}[u]] = \int_\Omega [2(u - f^\delta)\varphi + \varepsilon\varphi^2] d\mathbf{x} + \lambda \int_\Omega \frac{\nabla u \cdot \nabla\varphi}{\|\nabla u\|} d\mathbf{x} + O(\varepsilon^2)$$

and taking limit as $\varepsilon \to \infty$ we obtain the gradient or variational derivative for the ROF model

$$\langle \nabla\mathcal{J}[u]; \varphi \rangle = \int_\Omega 2(u - f^\delta)\varphi d\mathbf{x} + \lambda \int_\Omega \frac{\nabla u \cdot \nabla\varphi}{\|\nabla u\|} d\mathbf{x} \tag{5.17}$$

5.5 Total Variation

Now we use the divergence theorem $\int_\Omega \nabla \cdot \mathcal{F} = \int_{\partial\Omega} \mathcal{F} \cdot n$ and the formula $\nabla \cdot (\varphi \mathcal{F}) = \varphi \nabla \cdot \mathcal{F} + \nabla \varphi \cdot \mathcal{F}$ then

$$\int_\Omega [\varphi \nabla \cdot \mathcal{F} + \nabla \varphi \cdot \mathcal{F}] = \int_{\partial\Omega} \varphi \mathcal{F} \cdot n$$

to transform the second integral in (5.17) with

$$\mathcal{F} = \frac{\nabla u}{\|\nabla u\|}$$

$$\int \left(\nabla \cdot \frac{\nabla u}{\|\nabla u\|}\right)\varphi + \int \frac{\nabla \varphi \cdot \nabla u}{\|\nabla u\|} = \int \varphi \frac{\nabla u \cdot n}{\|\nabla u\|} = 0$$

then

$$\int \frac{\nabla \varphi \cdot \nabla u}{\|\nabla u\|} = -\int \left(\nabla \cdot \frac{\nabla u}{\|\nabla u\|}\right)\varphi$$

and

$$\langle \nabla \mathcal{J}[u]; \varphi \rangle = \int_\Omega \left(2(u - f^\delta) - \nabla \cdot \frac{\nabla u}{\|\nabla u\|}\right)\varphi$$

we finally have

$$\nabla \mathcal{J}[u] = 2(u - f^\delta) - \nabla \cdot \frac{\nabla u}{\|\nabla u\|}. \tag{5.18}$$

Observe that in the term

$$\nabla \cdot \frac{\nabla u}{\|\nabla u\|} = \frac{\partial}{\partial x}\left(\frac{u_x}{\sqrt{u_x^2 + u_y^2}}\right) + \frac{\partial}{\partial y}\left(\frac{u_y}{\sqrt{u_x^2 + u_y^2}}\right),$$

The presence of $\|\nabla u\|$ in the denominator causes computational difficulties when the magnitude of the gradient of the image becomes close to zero. One common remedy is to apply (5.15)

When ε gets close to zero, the smoothed energy becomes closer to the original total variation, but the optimization becomes more difficult. When ε becomes large, the smoothed energy becomes closer to the Sobolev energy, thus blurring the edges. Unfortunately, this prior is non-quadratic, and cannot be expressed over the Fourier domain. One thus need to use an iterative scheme such as a gradient descent to approximate the solution. An iteration of the gradient descent algorithm applied to the ROF model (5.16) is

Fig. 5.3 Observation with and without noise

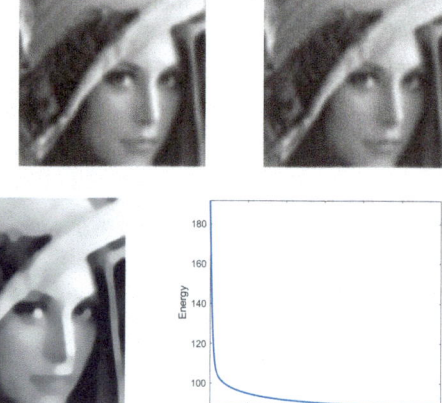

Fig. 5.4 (a) Result after applying TV REgularization by Gradient Descent applied to Fig. 5.3 (b) Convergence of the algorithm for obtaining (a)

$$u_{k+1} = u_k - \tau \left(h * (h * u_k - f^\delta) + \lambda \nabla \mathcal{J}(u_k) \right) \quad (5.19)$$

Example (TV Deconvolution) In this example we apply the Gradient algorithm [91] for solving the regularization problem given by the ROF model (5.16) (Figs. 5.3 and 5.4).

5.6 Convexity and Convex Functionals

Convexity and convex functionals play a crucial role in variational image processing due to their mathematical properties, which allow for more efficient and reliable optimization. Convex functionals are advantageous because (by Theorem 5.1) they guarantee that any local minimum is also a global minimum. This property simplifies the optimization process significantly. In fact, algorithms such as gradient descent, accelerated gradient methods, and Newton's method are particularly effective on convex functionals.

Convex functionals often lead to well-posed problems, meaning that solutions exist, are unique, and depend continuously on the data. This robustness is crucial for reliable image processing tasks.

Definition 5.2 A real-valued functional $\mathcal{J}$ defined on a convex subset C of a linear vector space $\mathcal{U}$ is said to be **convex** if

$$\mathcal{J}[\alpha u_1 + (1-\alpha)u_2] \leq \alpha \mathcal{J}[u_1] + (1-\alpha)\mathcal{J}[u_2] \quad (5.20)$$

5.6 Convexity and Convex Functionals

for all $u_1, u_2 \in C$ and all $\alpha \in]0, 1[$. If strict inequality ($<$) holds whenever $u_1 \neq u_2$, $\mathcal{J}$ is said to be **strictly convex**.

Example

1. The functions $f : \mathbb{R} \to \mathbb{R}; x \mapsto |x|^p$ are convex for $p \geq 1$
2. $f : \mathbb{R} \to \mathbb{R}$ in C^2 with $f''(x) \geq 0$ for all x then f is convex

The great utility of convexity lies in the fact that if to the properties of local extrema for nonlinear functionals we add the notion of convexity, these properties become global. In other words if $\mathcal{J}$ is a convex functional its local minima are global minima.

Theorem 5.1 *Let $\mathcal{J}$ be a convex functional defined on a convex subset C of a normed space. Let*

$$m = \inf_{u \in C} \mathcal{J}[u]$$

Then

1. *The subset $M = \{u \in C : \mathcal{J}[u] = m\}$ is convex*
2. *If u_0 is a local minimum of $\mathcal{J}$, then $\mathcal{J}[u_0] = m$ and, u_0 is a global minimum*

Proof

1. Consider $u_1, u_2 \in M$, then for $u = \alpha u_1 + (1 - \alpha)u_2 \; \alpha \in]0, 1]$, follows $\mathcal{J}[u] \leq \alpha \mathcal{J}[u_1] + (1 - \alpha)\mathcal{J}[u_2] = m$.
2. Consider a neighborhood $\mathcal{N}$ about u_0 which minimizes $\mathcal{J}$. If $u_1 \in C$, there exists $u \in M$ such that $u = \alpha u_0 + (1 - \alpha)u_1 \; \alpha \in]0, 1]$. Then $\mathcal{J}[u_0] \leq \mathcal{J}[u] \leq \alpha \mathcal{J}[u_0] + (1 - \alpha)\mathcal{J}[u_1]$ so $\mathcal{J}[u_0] \leq \mathcal{J}[u_1]$ □

Definition 5.3 *Let $\theta \geq 0$. A function $f : \mathbb{R}^N \to \mathbb{R}$ is θ-strongly convex if $u - \frac{\theta}{2}|x|$ is convex.*

If $u : \mathbb{R}^N \to \mathbb{R}$ is twice continuously differentiable, the following are equivalent [18]

- u is θ-strongly convex
- $\nabla^2 u \geq 0 \qquad \forall \mathbf{x} \in \mathbb{R}^N$
- $u(\mathbf{y}) \geq u(\mathbf{x}) + \nabla u(\mathbf{x}) \cdot (\mathbf{y} - \mathbf{x}) + \frac{\theta}{2}\|\mathbf{x} - \mathbf{y}\|^2 \qquad \forall \mathbf{x}, \mathbf{y} \in \mathbb{R}^N$
- $(\nabla u(\mathbf{x}) - \nabla u(\mathbf{y})) \cdot (\mathbf{x} - \mathbf{y}) \geq \theta|\mathbf{x} - \mathbf{y}^2| \qquad \forall \mathbf{x}, \mathbf{y} \in \mathbb{R}^N$

Some properties that may be useful for building convex functionals are the following. If $\mathcal{U}, \mathcal{V}$ are normed spaces and $\mathcal{J} : \mathcal{U} \to \mathbb{R}$ a convex functional, we have the following

1. *General properties*

    - For $\alpha \geq 0$ the functional $\alpha \mathcal{J}$ is convex
    - If $\mathcal{J}_1 : \mathcal{U} \to \mathbb{R}$ is convex, then so is $\mathcal{J} + \mathcal{J}_1$

- For $\phi : \mathbb{R} \to \mathbb{R}$ convex and monotonically increasing on the range of $\mathcal{J}$, the composition $\phi \circ \mathcal{J}$ is also convex.
- For $\Phi : \mathcal{V} \to \mathcal{U}$ affine linear, i.e.

$$\Phi(\alpha u + (1-\alpha)v) = \alpha \Phi(u) + (1-\alpha)\Phi(v) \qquad \forall u, v \in \mathcal{V}; \; \alpha \in [0, 1]$$

$\mathcal{J} \circ \Phi$ is convex on $\mathcal{V}$.

2. *Norms*

- Every norm $\|\cdot\|_{\mathcal{U}}$ on a normed space is convex, because

$$\|\alpha u + (1-\alpha)v\|_{\mathcal{U}} \leq |\alpha|\|u\|_{\mathcal{U}} + |1-\alpha|\|v\|_{\mathcal{U}}$$

but not strictly convex. Nevertheless the functional $\mathcal{J} = \phi(\|u\|_{\mathcal{U}})$ becomes strictly convex if $\phi[0, \infty[\to \mathbb{R}$ is strictly monotonically increasing and strictly convex and the norm in $\mathcal{U}$ is strictly convex [11]
- The norm in a Hilbert space is always strictly convex

3. *Integrals*

- Quadratic functionals

$$\mathcal{J}[u] = \int_{\Omega} \frac{1}{2} \nabla u(\mathbf{x})^T \mathbf{A}(\mathbf{x}) \nabla u(\mathbf{x}) + f(\mathbf{x}) u(\mathbf{x}) \tag{5.21}$$

are convex for a matrix (or operator) $\mathbf{A}$ symmetric and positive semidefinite
- Let $F : \bar{\Omega} \times \mathbb{R} \times \mathbb{R}^N$. Suppose $F(\mathbf{x}, y, \mathbf{z})$ is convex in $(y, \mathbf{z})$ $\forall \mathbf{x} \in \bar{\Omega}$, then under appropriate differentiability conditions (see [117]), the functional

$$\mathcal{J}[u] = \int_{\Omega} F(\mathbf{x}, u(\mathbf{x}), \nabla u(\mathbf{x})) d\mathbf{x} \tag{5.22}$$

is convex

4. *Gradient descent algorithm*[18]

For convex functionals, gradient descent algorithm

$$u_{k+1} = u_k - \alpha_k \nabla \mathcal{J}(u_k)$$

has desirable convergence properties.

- Since $\mathcal{J}(u)$ is convex, gradient descent is guaranteed to converge to the global minimum under appropriate conditions on the step size α_k.
- For a strongly convex functional, gradient descent converges at a linear rate, meaning that the error $\|u - u^*\|$ decreases exponentially with the number of iterations, where u^* is the global minimum.

5.6 Convexity and Convex Functionals

- Commonly, the step size can be chosen as a fixed constant, as a diminishing sequence α_k, or using line search methods. For strongly convex functionals, a fixed step size is often sufficient for achieving linear convergence.

Matlab Code

```
%Deconvolution by Total Variation Regularization
epsilon = 0.4*1e-2;
lambda = 0.06;
tau = 1.9 / ( 1 + lambda * 8 / epsilon );
fTV = y;
niter = 600;
Gr = grad(fTV);
d = sqrt( epsilon^2 + sum3(Gr.^2,3) );
G = -div( Gr./repmat(d, [1 1 2]) );
tv = sum(d(:));
e = Phi(fTV,h)-y;
fTV = fTV - tau*( Phi(e,h) + lambda*G);
%——————————————————————

tau = 1.9 / ( 1 + lambda * 8 / epsilon );
fTV = y;
E = [];
for i=1:niter
% Compute the gradient of the smoothed TV functional.
Gr = grad(fTV);
d = sqrt( epsilon^2 + sum3(Gr.^2,3) );
G = -div( Gr./repmat(d, [1 1 2]) );
% step
e = Phi(fTV,h)-y;
fTV = fTV - tau*( Phi(e,h) + lambda*G);
% energy
E(i) = 1/2*norm(e, 'fro')^2 + lambda*sum(d(:));
end
% display energy
figure
plot(E); axis('tight');
set_label('Iteration 3', 'Energy');

figure
imageplot(clamp(fTV));
```

Chapter 6
Variational Curve and Surface Reconstruction

This chapter delves into the principles and methods for reconstructing curves and surfaces from discrete data using variational techniques. We begin with an introduction to classical interpolation methods, focusing on polynomial Lagrange interpolation and its comparison with cubic spline interpolation. These approaches are evaluated for their smoothness and accuracy, particularly in the case of clean, structured data. The chapter then addresses the more challenging scenario of noisy data in one-dimensional cases, where simple interpolation fails to provide reliable reconstructions. To handle noise, we introduce variational regularization techniques, which stabilize the interpolation by balancing fidelity to the data and smoothness of the curve.

The discussion is extended to multivariate data and scattered data approximation, where classical methods become insufficient. In this context, we explore thin plate splines, a widely used tool for reconstructing smooth surfaces from scattered points. Regularization is incorporated into the spline framework to handle irregularities and noise in the data. Throughout the chapter, we provide theoretical insights and practical examples, illustrating how variational principles can be applied to reconstruct smooth, stable curves and surfaces from both structured and scattered data.

6.1 Polynomial Interpolation

Interpolation leverages approximation theory to model data. Essentially, approximation theory involves the art of representing a complex function using a simpler one, making it suitable for numerical implementation, typically on a computer. Although functions exist in infinite-dimensional spaces as abstract and ideal objects, algorithms must handle finite linear combinations of these functions.

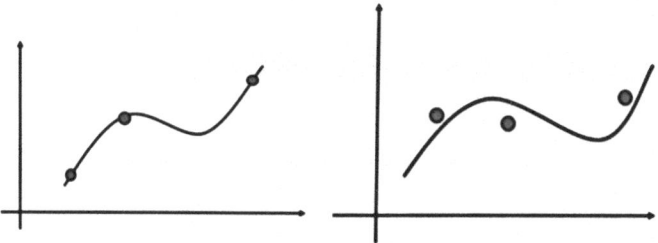

Fig. 6.1 Interpolation ($f(x_i) = y_i$) and approximation ($f(x_i) \approx y_i$)

In the univariate case, given a set of data $D = \{(x_i, y_i)\}_{i=0}^{M} \subset \mathbb{R}^2$, it is assumed there exist a functional relation f between the variables x and y such that $y_i = f(x_i)$ (Fig. 6.1). In the most simple case D represents two scalar variables whose values are obtained by sampling or experimentation; the **nodes**

$$a \leq x_0 < x_1 < \cdots < x_M \leq b \tag{6.1}$$

belong to an independent variable x and y_i depends on x_i.

The purpose of interpolation is to model this dependence making possible to estimate f at values that do not appear in data. The traditional and simplest method for solving the interpolation problem in one variable is to build a polynomial

$$P_M(x) = \sum_{i=0}^{M} \alpha_i x^i$$

of degree at most M, such that $y_i = P_M(x_i)$.

Polynomials are very common in computer science, so it is natural to ask why. There are several fundamental reasons for using them in the approximation of functions. From a purely algebraic point of view, they are the easiest functions to express and represent on a computer, calculating their derivatives and indefinite integrals is very simple and the results are also polynomials. Furthermore, there are many non-polynomial functions that can be expressed by power series that in the end also reduce to polynomials. The significance of polynomials is also supported by the Weierstrass approximation theorem: *For any function $f(x)$ continuous in $[a, b]$, there exists a sequence of ordinary polynomials which converges uniformly to $f(x)$ on $[a, b]$* [94].

On the other hand, if $x_0, x_1, \ldots, x_M$ are $M + 1$ distinct numbers and $y = f(x)$ a function with $y_j = f(x_j)$, then there exists a unique polynomial of degree at most such that $y_j = P_M(x_j)$ for $j = 0, 1, \ldots, M$. $P_M(x)$ is the classical **Lagrange polynomial**

$$P_M(x) = \sum_{k=0}^{M} y_k L_k(x), \quad \text{with } L_k(x) = \prod_{\substack{j=0 \\ j \neq k}}^{M} \frac{(x - x_j)}{(x_k - x_j)} \tag{6.2}$$

6.2 The Limitations of Polynomial Approximation

Table 6.1 Some terms of divided differences table

x_0	$f[x_0]$			
		$f[x_0, x_1] = \frac{f[x_1]-f[x_0]}{x_1-x_0}$		
x_1	$f[x_1]$		$f[x_0, x_1, x_2] = \frac{f[x_1,x_2]-f[x_0,x_1]}{x_2-x_0}$	$\ldots$
		$f[x_1, x_2] = \frac{f[x_2]-f[x_1]}{x_2-x_1}$		
x_2	$f[x_2]$			
$\vdots$				

For computational purpose is more convenient to use Newton's Divided-Difference form for Lagrange polynomial [13] written as

$$P_M(x) = f[x_0] + \sum_{k=1}^{M} f[x_0, x_1, \ldots, x_k](x - x_0) \ldots (x - x_{k-1}) \qquad (6.3)$$

where the $f[x_0, x_1, \ldots, x_k]$'s are obtained by the well-known divided differences method (Table 6.1).

Algorithm 6.1 Newton's divided differences

1: Input: $D = \{(x_i, f(x_i))\}_{i=1}^{M}$ such that $d_{i,0} = f(x_i)$
2: **for** $i = 1, 2, \ldots, M$ **do**
3:    **for** j=1,2,...,i **do**
4:       $d_{k,j} = \frac{d_{k,j-1} - d_{k-1,j-1}}{x_k - x_{k-j+1}}$;
5:    **end for**
6: **end for**
7: Output: $d_{i,i}, i = 0 \ldots M$ where

$$P_M(x) = d_{0,0} + \sum_{i=1}^{M} d_{i,i} \prod_{j=0}^{i-1}(x - x_j)$$

6.2 The Limitations of Polynomial Approximation

By the Weierstrass approximation theorem the quality of polynomial interpolation can be measured with

$$\|f - P_M\|_\infty = \max_{a \leq x \leq b} |f(x) - P_M(x)|$$

If the function f is continuous on $[a,b]$ one expect that P_M converge to f uniformly, that is $\|f - P_M\|_\infty \to \infty$ as M goes to infinity. However it can be proved (Faber's Theorem, [63]) that for any prescribed set of nodes $\{x_i\}$ (6.1), there exists a continuous function f on $[a,b]$ such that the interpolating polynomials for f using these nodes fail to converge uniformly to f.

Another common believe is that the larger the number of data points the better approximation. But while M increases, P_M also increase oscillations, then although P_M interpolates the M points, it also has a great distortion around these points (Fig. 6.3).

The limitations of classical interpolation indicate that additional information is necessary to achieve interpolants that better reflect the actual data. The spline functions, which we will explore in the next section, address the shortcomings of Lagrange polynomials by utilizing piecewise polynomial interpolation of a low degree.

6.3 Spline's Variational Theory

Splines are introduced as a solution to the rigidity of high-degree polynomials used in interpolation. While spline functions are also constructed from polynomials, they differ by being piecewise-defined. This construction retains the advantageous properties of polynomials while offering greater flexibility.

Definition 6.1 Let $a = x_0 < x_1 < \cdots < x_M = b$ be a subdivision of the interval $[a,b]$ and $m \in \mathbb{N}$. A function $S : [a,b] \to \mathbb{R}$ is called a spline of degree m with respect to this subdivision if (Fig. 6.2)

1. On each subinterval $[x_i, x_{i+1}]$ S is a polynomial of degree $\leq m$
2. $S, S', S'', ..., S^{(m-1)}$ are all continuous functions on $[a,b]$

This definition raises the question of what criteria should determine the most suitable spline degree. The answer lies in the variational interpretation of cubic polynomial splines. To reduce the oscillations associated with high-degree Lagrange polynomials, one effective approach is to minimize the integral curvature:

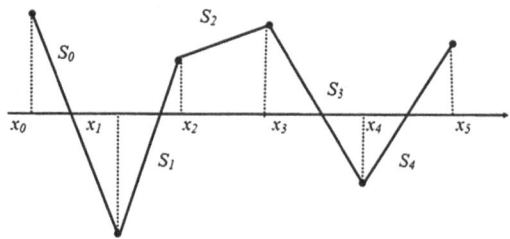

Fig. 6.2 Example for an order $m = 1$ spline

6.3 Spline's Variational Theory

$$\mathcal{J} = \int_a^b k^2(x)dx$$

Considering the well-known formula for the curvature

$$k(x) = \frac{y''}{(1+y'^2)^{3/2}} \qquad (6.4)$$

then

$$k^2 = \frac{(y'')^2}{(1+y'^2)^3} \approx (y'')^2$$

if the term y'^2 in the denominator (6.4) is small compared to 1 then

$$\mathcal{J}[y] = \int_a^b (y'')^2$$

is a good approximation. Assuming $y(x)$ sufficiently differentiable, calculating its first variation and applying integration by parts

$$\mathcal{J}[y+\varepsilon\varphi] = \int (y+\varepsilon\varphi)''^2 dx$$

$$= \int (y''^2 + 2\varepsilon y''\varphi'' + \varepsilon^2\varphi''^2)dx$$

$$\left.\frac{\partial \mathcal{J}[y+\varepsilon\varphi]}{\partial \varepsilon}\right|_{\varepsilon=0} = \int 2y''\varphi'' dx = 0$$

$$= -(-)\int y^{IV}\varphi dx = 0 \qquad \forall \varphi \in C_0^\infty$$

then, by fundamental lemma $y^{IV} = 0$. Solving this differential equations we obtain a cubic polynomial of the form $y(x) = a + bx + cx^2 + dx^3$.

This result shows the degree 3 in the interpolant polynomial as an optimal choice, in the sense that minimize a measure of the oscillations of the curve.

Definition 6.2 (Cubic Spline) Let $a = x_0 \leq x_1 \leq \cdots \leq x_M = b$ be a subdivision of the interval $[a, b]$ and $D = \{(x_j, y_j)\}_{j=0}^{M}$ are $M + 1$ points. The function $S(x)$ is called a cubic spline if there exist M cubic polynomials $S_j(x)$ with coefficients a_j, b_j, c_j, d_j such that

1. $S_j(x) = a_j + b_j(x - x_j) + c_j(x - x_j)^2 + d_j(x - x_j)^3$, $x \in [x_j, x_{j+1}]$, $j = 0, \ldots, M-1$
2. $S(x_j) = y_j \qquad j = 0, 1, \ldots, M$

3. $S_j(x_{j+1}) = S_{j+1}(x_{j+1})$ $j = 0, 1, ..., M - 2$
4. $S'_j(x_{j+1}) = S'_{j+1}(x_{j+1})$ $j = 0, 1, ..., M - 2$
5. $S''_j(x_{j+1}) = S''_{j+1}(x_{j+1})$ $j = 0, 1, ..., M - 2$

These five conditions are not enough for assure uniqueness, so two more conditions must be imposed: If $S''(x_0) = S''(x_M) = 0$, $S(x)$ is called *natural spline* and if $S'(x_0) = f'(x_0)$ and $S'(x_M) = f'(x_M)$ is called *clamped spline* [13, 78]. In this way the number of conditions equals the amount of coefficients for the spline. On each of the subintervals there is a different cubic polynomial $S_j(x)$ with four coefficients then we have to find $4M$ coefficients. Each $S_j(x)$ must satisfy two interpolation conditions on the ends of its corresponding interval, this gives $2M$ conditions. $S'_j(x)$ and $S''_j(x)$ must be continuous in the $M - 1$ interior points which gives $2(M - 1)$ conditions and finally for the natural spline second derivatives must vanish at 2 points. We finally have $2M + 2(M - 1) + 2 = 4M$ conditions equals to the number of coefficients.

Theorem 6.1 (Minimum Property of Cubic Splines) *Assuming $f \in C^2[a, b]$ and $S(x)$ the unique interpolant natural cubic spline for $f(x)$ such that $S''(x_0) = S''(x_M) = 0$ then*

$$\int_a^b [S''(x)]^2 dx \leq \int_a^b [f''(x)]^2 dx$$

Proof It's clear that

$$0 \leq \int_a^b (f'' - S'')^2 dx = \int_a^b (f''^2 - 2f''S'' + S''^2) dx$$

Now we calculate the second term $\int_a^b f''S'' dx$ in the former integral. Using integration by parts and the natural boundary condition $S''(x_0) = S''(x_M) = 0$ in the last expression we can see

$$\int_a^b S''(f'' - S'') dx = \left[S''(x)(f'(x) - S'(x))\right]_a^b - \int_a^b (f' - S') S''' dx$$

$$= 0 - \int_a^b (f' - S') S''' dx,$$

on every interval $[x_i, x_{i+1}]$ this last integral is zero because $S'''_j(x) = 6d_j$ and

$$\int_{x_i}^{x_{i+1}} (f' - S') dx = 6d_i \left[(f' - S')\right]_{x_i}^{x_{i+1}} = 0,$$

thus, $\int_a^b S''(x)(f''(x) - S''(x)) dx = 0$ and as a consequence $\int_a^b f'' S'' dx = \int_a^b S''^2 dx$ and replacing in (3.4)

6.3 Spline's Variational Theory

$$0 \leq \int_a^b (f'' - S'')^2 dx = \int_a^b (f''^2 - S''^2) dx$$

from which we finally obtain the expected result. □

Up to this point, we have discussed splines in their piecewise form. However, there are situations where a more compact representation is desirable. This can be achieved by embedding splines within the formal framework of linear spaces. By identifying well-defined, easily manageable sets of basis functions $U = \{u_1, u_2, \ldots, u_M\}$ we can approximate a function f by linear combinations sing linear combinations of these basis functions.

$$Pf = \sum_{j=1}^{M} \alpha_j u_j \tag{6.6}$$

We then measure the error in these approximations, necessitating at least the structure of a normed linear space. This framework allows us to transition from the piecewise definition of splines, which is convenient for computer representation, to more compact forms useful for theoretical purposes. The basis functions selected for this process are

$$(x - t)_+^n = \begin{cases} (x - t)^n, & x \geq t \\ 0, & x < t \end{cases} \tag{6.7}$$

Its well-known that $\{1, x, x^2, \ldots, x^k, (x - x_0)_+^k, (x - x_1)_+^k, \ldots, (x - x_{M-1})_+^k\}$ form a basis for the space of splines of order k [63]. Thus every cubic spline has a representation of the form

$$S(x) = \sum_{j=1}^{M} \alpha_j (x - x_j)_+^3 + \beta_1 + \beta_2 x \tag{6.8}$$

and given the set $a = x_0 < x_1 < \ldots < x_M = b$ there is a unique natural cubic spline taking defined on these knots.

The form (6.8) is not well conditioned for computational work, however can be translated to a more compact form using translates of the form $|x - x_i|^3$, as in the next result.

Theorem 6.2 *Given* $a = x_1 < x_2 < \cdots < x_M = b$, *every natural cubic interpolating spline* $S(x)$ *has a representation of the form*

$$S(x) = \sum_{j=1}^{M} \alpha_j |x - x_j|^3 + \beta_1 + \beta_2 x, \quad x \in \mathbb{R} \tag{6.9}$$

The coefficients $\{\alpha_j\}$ have to satisfy

$$\sum_{j=1}^{M} \alpha_j = \sum_{j=1}^{M} \alpha_j x_j = 0 \qquad (6.10)$$

Given the data $D = \{(x_j, y_j)\}_{j=1}^{M}$, in order to determine the α_j's is necessary to solve the system

$$\begin{bmatrix} |x_1-x_1|^3 & |x_1-x_2|^3 & \cdots & |x_1-x_M|^3 & 1 & x_1 \\ |x_1-x_1|^3 & |x_1-x_1|^3 & \cdots & |x_1-x_1|^3 & 1 & x_2 \\ \vdots & \vdots & \ddots & \vdots & \vdots & \vdots \\ |x_M-x_1|^3 & |x_M-x_2|^3 & \cdots & |x_M-x_M|^3 & 1 & x_M \\ 1 & 1 & \cdots & 1 & 0 & 0 \\ x_1 & x_2 & \cdots & x_M & 0 & 0 \end{bmatrix} \begin{bmatrix} \alpha_1 \\ \alpha_2 \\ \vdots \\ \alpha_M \\ \beta_1 \\ \beta_2 \end{bmatrix} = \begin{bmatrix} y_1 \\ y_2 \\ \vdots \\ y_M \\ 0 \\ 0 \end{bmatrix} \qquad (6.11)$$

$$\underbrace{}_{Interpolation\ Matrix}$$

where the rows in the above matrix correspond to the interpolation conditions in (6.9) and the last two rows are the application of (6.10). If we define

$$P^T = \begin{bmatrix} 1 & 1 & \cdots & 1 \\ x_1 & x_2 & \cdots & x_M \end{bmatrix}, \qquad A_{i,j} = |x_i - x_j|^3$$

the system can be written as

$$\begin{bmatrix} A & P \\ P^T & 0 \end{bmatrix} \begin{bmatrix} \alpha \\ \beta \end{bmatrix} = \begin{bmatrix} y \\ 0 \end{bmatrix} \qquad (6.12)$$

with

$$\boldsymbol{\alpha} = [\alpha_1, \ldots, \alpha_M]^T, \boldsymbol{\beta} = [\beta_1, \beta_2]^T, \mathbf{y} = [y_1, \ldots, y_M]^T$$

and the matrix of the system (6.12) is called **the interpolation matrix**. This result is demonstrated in the following section as a particular case of smoothing splines. By the moment in the next example we illustrate the power of this representation and its accuracy.

Example Figure 6.3 shows interpolation of exact data points (without noise) located on the Runge's function

$$f(x) = \frac{1}{1+x^2},$$

comparing Lagrange polynomials and splines. The Lagrange interpolation is done with the Newton's form (6.3) and splines with (6.9) (see the Matlab code at the

6.4 Regularization and Smoothing Splines

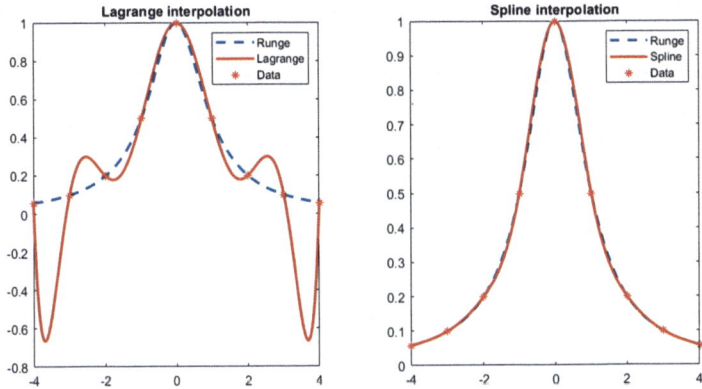

Fig. 6.3 Comparing interpolation by splines and Lagrange polynomials

end of this chapter). Left figure shows the big limitations of Lagrange polynomials which is overcome by spline interpolations (right figure). □

6.4 Regularization and Smoothing Splines

In many real problems, data does not appear in exact form, frequently they are noisy. Noise is usually considered as an undesired perturbation and appears during every data acquisition process. In this case is not advisable to interpolate data ($y_i = f(x_i)$) but to approximate them ($y_i \approx f(x_i)$). Thus is better to apply the so called smoothing splines, which unlike interpolating splines, may not contain the data points (x_i, y_i). This problem is better solved by approaching the situation as an inverse problem that can be solved by minimization of the functional.

$$\mathscr{T}[f] = \underbrace{\sum_{j=1}^{M}[f(x_j) - y_j]^2}_{\text{Fidelity to data}} + \lambda \overbrace{\int (f''(x))^2 dx}^{\text{Generalization}} \tag{6.13}$$

whose solution is exactly a natural cubic spline. The cost functional in (6.13) has two parts: $E[f] = \sum_{j=1}^{M}(f(x_j) - y_j)^2$, that is minimized by an straight line and $\mathcal{R}[f] = \int f''^2(x)dx$, the regularization functional, is minimized by a cubic spline.

The parameter λ trades off the importance of these two competing costs in (6.13). For small λ, the minimizer is close to an interpolating spline. For λ large, the minimizer is closer to a straight line. In statistics this method is sometimes called penalized squares and is seen rather as a form of nonparametric regression; a form of regression analysis in which the predictor does not take a predetermined form but is constructed according to information derived from the data; its goal is to construct a model for f and estimate it based on noisy data. In the simple case when both the dependent variable and the independent variable are scalar variables, one type of regression model relates dependent and independent variables as

$$y_i = f(x_i) + \varepsilon_i, \qquad i = 1, ..., M$$

where f is the regression function and ε_i are zero-mean independent random errors with common variance σ^2.

Theorem 6.3 ([122]) *The solution to the optimization problem (6.13) is a natural cubic spline which has a representation of the form*

$$S(x) = \sum_{j=1}^{M} \alpha_j \phi(|x - x_j|) + p(x), x \in \mathbb{R} \tag{6.14}$$

where $\phi(r) = r^3, r \geq 0$, and $p \in \Pi_1(\mathbb{R})$. The coefficients α_j have to satisfy (6.10). In the same way, for every set $\mathcal{A} = \{x_1, x_2, \ldots, x_M\}$ of pairwise distinct points and for every $\mathbf{y} \in \mathbb{R}^N$, there exists a function S of the form (6.22) with (6.10) that interpolates the data, this is $S(x_j) = y_j, 1 \leq j \leq M$.

Applying the properties of Dirac's distribution, $f(x_i) = \langle \delta_{x_i}, f \rangle = \int f(x)\delta(x - x_i)dx$, the functional (6.13) can be written as

$$\mathcal{T}[f] = \sum_{j=1}^{M} (f(x_j) - y_j)^2 + \lambda \int f''^2 dx$$

$$= \sum_{j=1}^{M} \int (f(x) - y_j)^2 \delta_{(x_j)} + \lambda \int f''^2 dx$$

$$= \int \sum_{j=1}^{M} (f(x) - y_j)^2 \delta_{(x_j)} + \lambda \int f''^2 dx$$

Calculating the functional gradient $\nabla \mathcal{T}[f]$

$$\nabla \mathcal{T} = 2 \sum_{j=1}^{M} (f(x) - y_j)\delta_{(x_j)} + \lambda \frac{d^4 f}{dx^4} = 0$$

$$\frac{d^4 f}{dx^4} = \frac{1}{\lambda} \sum_{j=1}^{M} (f(x) - y_j)\delta_{(x_j)}$$

$$= \sum_{j=1}^{M} \frac{(y_j - f(x))}{\lambda} \delta_{(x_j)}$$

6.4 Regularization and Smoothing Splines

The solution $S(x)$ to this differential equation can be written as the convolution of the source term with the fundamental solution of the operator $f^{IV} = \delta$. So we calculate the Green's function $K(x, \xi)$ such that

$$\frac{d^4 K(x, \xi)}{dx^4} = \delta(x - \xi)$$

then by (3.39), $K(x, \xi) = |x - \xi|^3$ and

$$S(x) = K * \sum_{j=1}^{M} \frac{y_j - f(x)}{\lambda} \delta(x - x_j)$$

Defining

$$\alpha_j = \frac{y_j - f(x_j)}{\lambda}, \qquad (6.15)$$

we have the spline solution

$$S(x) = \sum_{j=1}^{M} \alpha_j |x - x_j|^3$$

As the null space of $\mathcal{R}[f]$ is $\Pi_1(\mathbb{R})$, the polynomials of degree less or equal than one, we have to add a term $p(x) \in \Pi_1(\mathbb{R})$ written as a linear combination of any base for $\Pi_1(\mathbb{R})$, in particular we can use $\{1, x\}$. The final expression for the solution of the optimization problem is

$$S_\lambda(x) = \sum_{j=1}^{M} \alpha_j \phi(|x - x_j|) + p(x) = \sum_{j=1}^{M} \alpha_j |x - x_j|^3 + \beta_1 + \beta_2 x \qquad (6.16)$$

In order to find the values for α_i we translate the problem to matrix form. Using (6.15) we can write $y_i = f(x_i) + \lambda \alpha_i$, then

$$y_i = \sum_{j=1}^{M} \alpha_j |x_i - x| + \beta_1 + \beta_2 x_i + \lambda \alpha_i, \quad i = 1, 2, ..., M$$

If this is written in matrix form $\mathbf{y} = A\boldsymbol{\alpha} + P\boldsymbol{\beta} + \lambda\boldsymbol{\alpha}$, we obtain

$$\begin{cases} (A + \lambda I)\boldsymbol{\alpha} + P\boldsymbol{\beta} = \mathbf{y} \\ P^T \boldsymbol{\alpha} = 0 \end{cases}$$

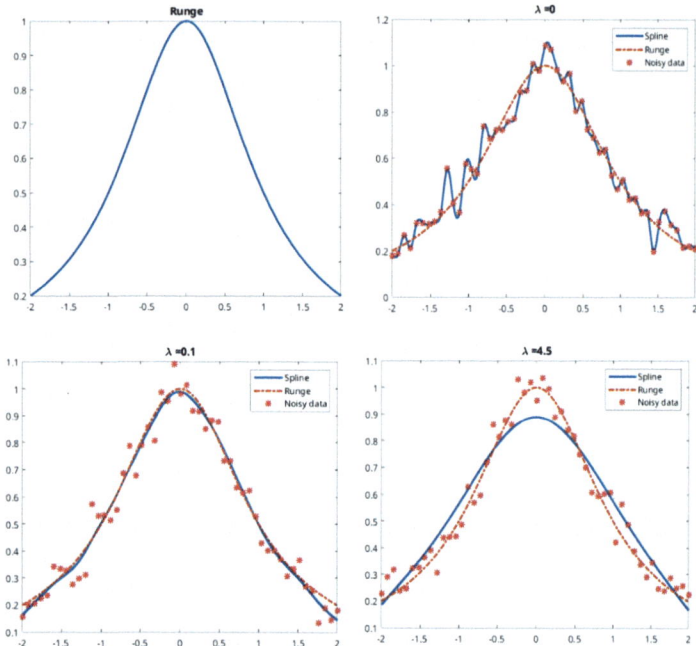

Fig. 6.4 Spline smoothing with different values for λ

which in matrix form is

$$\begin{bmatrix} A + \lambda I & P \\ P^T & 0 \end{bmatrix} \begin{bmatrix} \alpha \\ \beta \end{bmatrix} = \begin{bmatrix} y \\ 0 \end{bmatrix} \tag{6.17}$$

Example Figure 6.4 shows the results of smoothing noisy data on the Runge function with different values of the regularization parameter λ. For $\lambda = 0$ the spline just reproduce the noise (overfitting); $\lambda = 0.1$ is a good trade off between generalization and fidelity to data; $\lambda = 4.5$ is too large (underfitting). □

6.5 Extensions to Higher Dimensions

Historically after the application of variational approach to splines theory arouse a great interest in application and generalizations of these techniques, especially their extensions to several variables. In fact, cubic spline is a particular case of the problem considered by Schoenberg [106], about finding and interpolator that minimize the functional $J[f] = \int_a^b (f^m(x))^2 dx$.

6.5 Extensions to Higher Dimensions

During the 1970s, Wahba and others demonstrated the connection between smoothing splines and Bayesian estimates, highlighting their strong approximation and theoretical properties along with various generalizations. Within this framework, f can be either a fixed function with a certain degree of smoothness or a sample function from a stochastic process. With the advancement of computer capabilities, these results became practical and attractive for numerous applications.

The development of interpolation and approximation methods from one variable to multiple variables has taken various forms, such as tensor products, piecewise bivariate polynomials, and box splines. These methods often aim to extend the properties of cubic splines to higher dimensions. However, the most effective approach to generalizing splines and smoothing splines to multiple variables has been through their variational characterization. This was achieved by Duchon [34], Meinguet [79], and others, resulting in the formulation of thin plate splines (TPS).

Besides its theoretical properties and physical interpretations, it was the representation as radial basis function (RBF) which gave it success in practical implementations. A RBF is a linear combination of translates of a fixed radial function. The function ϕ is called radial because it is the composition of a univariate function with the Euclidian norm $\|\cdot\|_2$ on $\mathbb{R}^N$. A straightforward generalization is to build interpolants of the form

$$S(x) = \sum_{j=1}^{M} \alpha_j \phi(\|x - x_j\|_2) + p(x) \quad x \in \mathbb{R}, \tag{6.18}$$

where $\phi : [0, \infty[\to \mathbb{R}$ is a univariate fixed function and $p \in \Pi_{m-1}(\mathbb{R})$ is a low degree n-variate polynomial. The additional conditions on the coefficients now become

$$\sum_{j=1}^{M} \alpha_j q(x_j) = 0 \quad \forall q \in \Pi_{m-1}(\mathbb{R}) \tag{6.19}$$

In many cases it is not necessary the additional polynomial $p(x)$ and its corresponding condition (6.19). In this particular case the interpolation problem reduce to the question whether the matrix $A = [\phi(\|x_i - x_j\|)]$ is nonsingular. In other words, does there exists a function ϕ such that for all pairwise distinct points $\mathcal{A} = \{x_1, x_2, \ldots, x_M\} \subset \mathbb{R}$ the matrix A is nonsingular? The answer is yes. Examples are the Gaussian $\phi(r) = e^{-cr^2}$ and the multiquadric. $\phi(r) = \sqrt{c^2 + r^2}, c > 0$. Using these ideas, the properties of cubic splines can be generalized to the multivariate case.

6.6 Thin Plate Spline and Scattered Data Interpolation

Interpolation of scattered data [39, 79] is a fundamental problem in numerical analysis and computational mathematics. It involves constructing a smooth surface or function that approximates data points irregularly distributed in space. Scattered data points (Fig. 6.5) do not follow a regular grid, making it challenging to apply traditional interpolation methods designed for structured data.

This problem is common in various fields such as geostatistics, computer graphics, medical imaging, and machine learning. Among the various methods for achieving this, Thin Plate Splines have gained prominence due to their flexibility and ability to produce smooth interpolations.

TPS is a generalization of cubic splines to multiple dimensions and is based on the minimization of a variational problem. Its name comes from the analogy of bending a thin metal sheet, which seeks the minimal bending energy.

The interpolation problem to scattered data can be formulated as follows: Given a finite set $\mathcal{A} = \{\mathbf{x}_1, \mathbf{x}_2, \ldots, \mathbf{x}_M\}$ of distinct points of $\mathbb{R}^N$ and associate (real) values $\{y_i\}_{i \in I}$, construct a (continuous) function

$$f : \mathbb{R}^N \to \mathbb{R} \qquad \text{with} \qquad f(\mathbf{x}_i) = y_i; \qquad i \in I$$

and minimize the seminorm

$$|f|_m^2 = \int_{\mathbb{R}^N} \sum_{|\alpha|=m} c_\alpha |\partial^\alpha f(\mathbf{x})|^2 d\mathbf{x} \tag{6.20}$$

The kernel of this seminorm is the linear space $\Pi_m(\mathbb{R}^N)$ of polynomials in N variables of (total) degree $\leq m - 1$, with dimension $\binom{m+N-1}{n}$.

For the case $m = N = 2$, the seminorm $|f|_m^2$ is physically interpreted as the energy of a thin plate of infinite extension, where f is the deflection perpendicular to the rest position (which is supposed to be a plane). This makes an analogy with

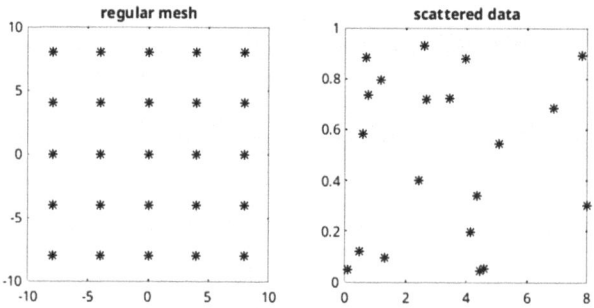

Fig. 6.5 Thin plate spline solves the problem of interpolation on scattered data

6.6 Thin Plate Spline and Scattered Data Interpolation

the cubic spline, which minimizes the potential energy of a statically deflected thin beam. We then have the thin plate energy (6.20) for the case $m = 2$ in $\mathbb{R}^2$

$$\mathcal{R}[f] = |f|_2^2 = \int_{\mathbb{R}^2} \left(\frac{\partial^2 f}{\partial x^2}\right)^2 + 2\left(\frac{\partial^2 f}{\partial x \partial y}\right)^2 + \left(\frac{\partial^2 f}{\partial y^2}\right)^2 \tag{6.21}$$

So applying the scheme (5.6) the interpolation problem in $\mathbb{R}^2$ is solved by minimizing the functional

$$\mathcal{T}[f] = \sum_{j=1}^{M} [f(\mathbf{x}_j) - y_j]^2 + \lambda \int_{\mathbb{R}^2} \left(\frac{\partial^2 f}{\partial x^2}\right)^2 + 2\left(\frac{\partial^2 f}{\partial x \partial y}\right)^2 + \left(\frac{\partial^2 f}{\partial y^2}\right)^2 \tag{6.22}$$

thus, applying properties of distributions

$$\mathcal{T} = \int_{\mathbb{R}^2} \sum_j \left(f(\mathbf{x}) - y_j\right)^2 \delta_{(\mathbf{x}_j)} + \lambda \int_{\mathbb{R}^2} \left(\frac{\partial^2 f}{\partial x^2}\right)^2 + 2\left(\frac{\partial^2 f}{\partial x \partial y}\right)^2 + \left(\frac{\partial^2 f}{\partial y^2}\right)^2$$

so applying Euler-Lagrange equation (4.17) for $\mathcal{T}$

$$2\sum_j \left(f(\mathbf{x}) - \mathbf{y}_j\right)^2 \delta_{(\mathbf{x}_j)} + 2\lambda \left(\frac{\partial^4 f}{\partial x^4} + \frac{\partial^4 f}{\partial x^2 \partial y^2} + \frac{\partial^4 f}{\partial y^4}\right) = 0$$

simplifying and using (4.19) for the biharmonic operator

$$\sum_j \left(f(\mathbf{x}) - \mathbf{y}_j\right)^2 \delta(\mathbf{x} - \mathbf{x}_j) + \lambda \Delta^2 f = 0$$

We then have the differential equation

$$\Delta^2 f = \sum_{j=1}^{M} \frac{(y_j - f(\mathbf{x}))}{\lambda} \delta(\mathbf{x} - \mathbf{x}_j)$$

whose solution by (3.38) is

$$f(\mathbf{x}) = K * \sum_{j=1}^{M} \frac{(y_j - f(\mathbf{x}))}{\lambda} \delta(\mathbf{x} - \mathbf{x}_j)$$

and K is the fundamental solution (3.42) of the biharmonic operator. Denoting $\alpha_j := (y_j - f(\mathbf{x}))/\lambda$

$$f(\mathbf{x}) = K * \sum_j \alpha_j \delta(\mathbf{x} - \mathbf{x}_j) = \sum_j \alpha_j K * \delta(\mathbf{x} - \mathbf{x}_j)$$

Fig. 6.6 Franke's function in $[0, 1] \times [0, 1]$

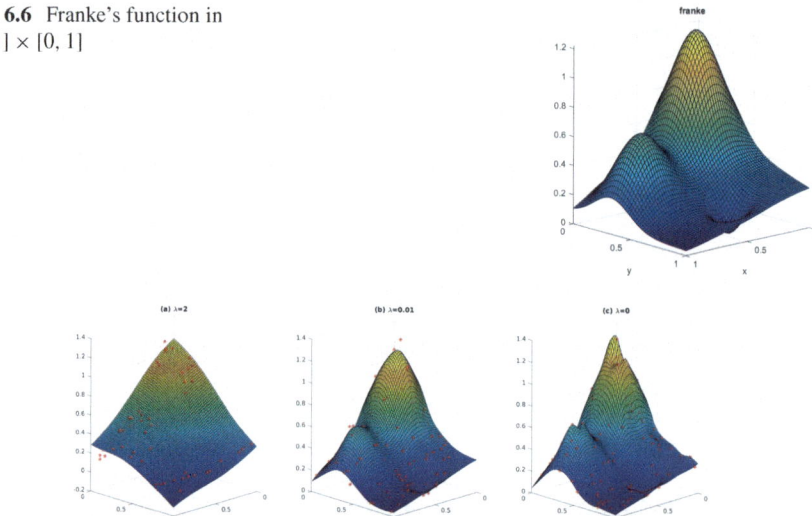

Fig. 6.7 Smoothing of scattered noisy data by splines. (**a**) λ too large (underfitting) (**b**) λ adequate, (**c**) $\lambda = 0$, interpolates the noise (overfitting)

and applying the property (3.26) for convolutions

$$f = \sum_j \alpha_j K(\mathbf{x} - \mathbf{x}_j) \tag{6.23}$$

Since the kernel of (6.21) is $\Pi_1(\mathbb{R}^2)$ we have to add a polynomial $p(\mathbf{x}) \in \Pi_1(\mathbb{R}^2)$ to the final expression for the solution, so

$$S(\mathbf{x}) = \sum_j \alpha_j K(\mathbf{x} - \mathbf{x}_j) + p(\mathbf{x}) = \sum_{j=1}^{M} \alpha_j \phi(\|\mathbf{x} - \mathbf{x}_j\|) + p(\mathbf{x}) \tag{6.24}$$

where $\phi(r) = r^2 \ln(r)$ is the radial basis function corresponding to thin plate spline.

Example Figure 6.7 shows the results for the reconstruction of Franke's function (Fig. 6.6) from scattered noisy data applying different values of λ, the regularization parameter

$$\text{franke}(x, y) = \frac{3}{4} e^{-\frac{(9x-2)^2}{4} - \frac{(9y-2)^2}{4}} + \frac{3}{4} e^{-\frac{(9x+1)^2}{49} - \frac{(9y+1)}{10}} + \frac{1}{2} e^{-\frac{(9x-7)^2}{4} - \frac{(9y-3)^2}{4}} - \frac{1}{5} e^{-(9x-4)^2 - (9y-7)^2}$$

6.6 Thin Plate Spline and Scattered Data Interpolation

There exists an extensive literature on radial basis functions [39] and algorithms for estimating optimal values of λ [122], but that is beyond the scope of this book.

□

Listing 6.1 Spline functions

```
%Example for interpolation: >>comparar(4,8,0,0)
%Example for Smoothing:>>comparar(2,50,.03,0.01)

%-1------------------main function------------------


function comparar(a,N,ruido,lambda)
%Draw graphs for Runge, Lagrange and spline functions
%a: length of the interval
%N: Points number
%ruido: noise level
%lambda: regularization parameter
%for interpolation of exact data Ruido=0;lambda=0
clc;close all;

runge=@(x) 1./(1+x.^2);
xknots=(-a:2*a/N:a)' ;
yruido=runge(xknots) + ruido*randn(length(xknots),1);
[c,d]=newpol(xknots,yruido);
t=-a:0.01:a;
yexacto=runge(t);                      %Runge
lagrange=fnewton(c,xknots,t);     %Lagrange
L=splinecubico(xknots,yruido,lambda);
splinef=splcubic(L,xknots,t);     %cubic spline

%------------------INTERPOLATION------------------
figure
subplot(1,2,1)
plot(t,yexacto,'--','LineWidth',2),hold on %Runge +Lagrange
plot(t,lagrange,'LineWidth',2)
plot(xknots,yruido,'*r'),title('Lagrange interpolation ')
legend('Runge','Lagrange','Data')

subplot(1,2,2)
plot(t,yexacto,'--','LineWidth',2),hold on
plot(t,splinef,'LineWidth',2)
plot(xknots,yruido,'*r')
title('Spline interpolation ')
legend('Runge','Spline ','Data')
%------------------

% ------------------SMOOTHING------------------
% figure
% plot(t,splinef,'LineWidth',2)
% hold on
```

```
% plot(t,yexacto,'-.','LineWidth',2)
% plot(xknots,yruido,'*r')
% legend('Spline', 'Runge','Noisy data')
%-

end

%-
%Example for interpolation: >>comparar(4,8,0,0)
%Example for Smoothing:>>comparar(2,50,.03,0.01)


%-2

function alfa=splinecubico(td,y,lambda)
%Obtains the weights for spline
%td x-coordinates of knots
% y: data to be interpolated
% alfa:weight vector
A = matrizcubica(td,lambda);
b= [y;0 ;0];
alfa= A\b;


%-3

function val= splcubic(alfa,td,x)
%Evaluate spline in x
%alfa: column with the weights of the spline
%td column of knots
%x evaluation vector

M=length(x);
N=length(td);
ax = x'*ones(1,N);
at=ones(M,1)*td';
A = abs(ax-at).^3;

val = A*alfa(1:N)+ alfa(N+1)*ones(M,1)+ alfa(N+2)*x';

end


%-4

function A =  matrizcubica(td,lambda)
%td: x-coordinates of the knots
M =length(td);

unos=ones(M,1);
A1 = td*unos';
B =(abs(A1-A1')).^3;
```

```
P =[unos td];
A= [ B+lambda*eye(M) P;P' [0 0;0 0]];
```

%—5————————————————————

```
%Call as [c,d]=newpol(x,y)
%x,y: row data vectors
%d:divided differences matrix
%c:diagonal of d for f[xo..xN] in Newton form
function [c,d]=newpol(x,y)
N =length(x);
d=zeros(N,N);
d(:,1)= y';

for j=2:N
for k=j:N
d(k,j)=(d(k,j-1)-d(k-1,j-1))/(x(k)-x(k-j+1));
end
end
c=diag(d);
```

%—6————————————————————

```
function w = fnewton(c,xdata,x)
%call as w = fnewton(c,xdata,x)

%Evaluate Newton's Lagrange form on a vector x
% x: vector to evaluate the polinomial
% c diagonal of divided differences matrix
%xdata :x-coordinates of the knots
N= length(xdata);
M=length(x);
w=zeros(M,1);
for j=1:M
w(j) =c'*cumprod([1;x(j).*ones(N-1,1)-xdata(1:N-1)]);
end
```

%————————————————————————

Chapter 7
Variational Image Feature Extraction

Variational image feature extraction refers to using techniques from variational calculus to extract meaningful image features such as edges, textures, and shapes by minimizing energy functionals. This approach offers a powerful framework for image processing tasks like denoising, segmentation, and deblurring while respecting important properties such as regularity, smoothness, and sharp transitions.

We begin with the Mumford-Shah functional, a foundational approach for partitioning an image into meaningful regions while preserving edges. This functional is central to many modern variational methods and is particularly well-suited for capturing the trade-off between smoothness and edge preservation.

Next, we delve into Variational Level Set Methods and Active Contours, which provide flexible frameworks for tracking evolving shapes and interfaces. These methods use implicit representations of contours, making them powerful tools for problems involving topological changes, such as merging or splitting objects.

Finally, we introduce the Chan-Vese model, a specialized case of the Mumford-Shah functional that simplifies image segmentation by focusing on piecewise constant regions. This model has gained wide popularity for its efficiency and robustness in segmenting images with intensity inhomogeneities and noise.

Throughout the chapter, theoretical insights are paired with practical examples to illustrate the application of these variational models to real-world image segmentation challenges [76, 80, 81, 87].

7.1 The Mumford-Shah Functional

In mathematics, *free discontinuity problems* often refer to a class of variational problems that involve finding a solution with prescribed discontinuities in certain regions of the domain. These problems are frequently encountered in image processing and computer vision, where the goal is to model and solve problems with discontinuities or sharp transitions.

One notable framework associated with free discontinuity problems is the Mumford-Shah model, which is a variational approach to image segmentation and denoising. The paper by David Mumford and Jayant Shah [83], 1989, is a seminal work in the field of image processing and computational mathematics. Since their model was introduced, it has been extensively examined and refined by numerous researchers. This work includes studying the characteristics of minimizers, developing various approximations, and applying the model to a wide range of tasks such as image segmentation, partitioning, and restoration, as well as broader applications in image analysis and computer vision.

The Mumford-Shah model provides a framework for image segmentation while simultaneously handling the challenge of discontinuities (such as edges) and noise in images. This is to decompose an image into regions with different intensity values and characterized by the presence of free (unconstrained) discontinuities.

If $(g_{i,j})$ is a noisy image coming from $(f_{i,j})$ the discrete problem, as presented by [47], suggests looking for f as a minimizer of the "weak membrane" energy

$$E(f) = \sum_{i,j} W(|f_{i+1,j} - f_{i,j}|) + W(|f_{i,j+1} - f_{i,j}|) + |f_{i,j} - g_{i,j}|^2$$

In this energy the terms where function W appears are regularity terms. The other term, $\sum_{i,j} |f_{i,j} - g_{i,j}|^2$, ensures that f remains close to the original data g. This problem is equivalent to minimizing Geman's energy [8]:

$$E(f) = \sum_{i,j} \{\lambda^2 |f_{i+1,j} - f_{i,j}|^2 (1 - v_{i,j}) + \lambda^2 |f_{i,j+1} - f_{i,j}|^2 (1 - h_{i,j}) \\ + \alpha(h_{i,j} + v_{i,j}) + |f_{i,j} - g_{i,j}|^2\} \quad (7.1)$$

Inspired in this kind of reasoning, in their original paper model Mumford and Shah propose to minimize the functional

$$E(f, \Gamma) = \mu^2 \iint_R (f - g)^2 dy dx + \iint_{R\setminus\Gamma} \|\nabla f\|^2 dy dx + v|\Gamma| \quad (7.2)$$

where $|\Gamma|$ stands for the total length of the arcs making up Γ. The function f represents a smooth approximation of the given image g, whereas the set Γ is a set of curves coming from the sharp discontinuities of g.

To make a better analysis of the properties of the Mumford-Shah functional, it is customary to express it using the Hausdorff measure $\mathcal{H}^n(K)$. For $K \subset \mathbb{R}^N$ and $n > 0$

$$\mathcal{H}^n(K) := \sup_{\varepsilon > 0} \mathcal{H}^n_\varepsilon(K)$$

7.1 The Mumford-Shah Functional

called the n-dimensional Hausdorff measure of the set K, where

$$\mathcal{H}^n_\varepsilon(K) = C_n \inf \left(\sum_i (diam\, A_i)^n \right)$$

taken over all countable families $\{A_i\}_{i=1}^\infty$ of open sets A_i such that

$$K \subset \bigcup_{i=1}^\infty A_i; \qquad diam\, A_i \leq \varepsilon \quad \forall i$$

C_n is a constant such that $\mathcal{H}^n$ reduces to Lebesgue measure on planes of dimension n. So the standard Mumford-Shah model is

$$\mathcal{T}^{MS}_{\alpha,\beta,u^\delta}(u, K) = \int_\Omega (u - u^\delta)^2 + \alpha \int_{\Omega \setminus K} |\nabla u|^2 + \beta \mathcal{H}^{N-1}(K) \qquad (7.3)$$

with regularization parameters $\alpha, \beta > 0$; $\Omega \subset \mathbb{R}^N$ is the image domain, u^δ is the recorded image. $\mathcal{H}^{N-1}(K)$ is the "size" or Hausdorff measure and stands for the total surface measure of the hypersurface K (the counting measure if $N = 1$, the length measure if $N = 2$, the area measure if $N = 3$). So the segmentation of the image u^δ is a pair (u, K) that minimize the Mumford–Shah functional such that u is piecewise smooth function with discontinuities along the set K, which coincides with the object boundaries

- the fidelity term $\int_\Omega (u - u^\delta)^2 d\mathbf{x}$ is forcing the solution u to be as close as possible to the given image u^δ. It promotes the fidelity of the segmented image to the original data.
- the smoothing term $\int_{\Omega \setminus K} |\nabla u|^2 d\mathbf{x}$ forces the solution u to be as smooth as possible everywhere except along the image discontinuities;
- the geometric term $\mathcal{H}^{N-1}(K)$ forces the total length of the edges to be as short as possible. This encourages the algorithm to find regions with a relatively uniform intensity. It helps in creating coherent segments in the image.

The segmentation problem in image analysis and computer vision consists in computing a decomposition where Ω_i are disjoint connected open subsets of Ω each one with a piece-wise smooth boundary and K is the union of the part of the boundaries of the Ω_i inside Ω, such that $\Omega = \bigcup \Omega_i \cup K$ and

- The image u^δ varies smoothly and/or slowly within each Ω_i.
- The image u^δ varies discontinuously and/or rapidly across most of the boundary K between different Ω_i.

If in (7.3) the function u is constrained to the form

$$\{u : u = a_i\, on\, \Omega_i; a_i \in \mathbb{R}\}$$

the functional becomes the *piecewise constant Mumford-Shah functional*

$$\mathscr{T}_{MS}(u, K) = \sum_i \int_{\Omega_i} (u^\delta - a_i)^2 + \beta \mathcal{H}^1(K) \tag{7.4}$$

and the minimizing function is $u = \sum_i a_i 1_{\Omega_i}$ with $mean_{\Omega_i} := \int_{\Omega_i} u^\delta$ and for the solution u, a_i are the mean values of u^δ on Ω_i. If K is fixed in (7.3), then for a sequence $\alpha_j \to \infty$ the sequence of minimizers u_j of $\mathscr{J}_{\alpha,\beta,u^\delta}$ tends to a piecewise constant function. So (7.4) is a limit functional of (7.3) when $\alpha_j \to \infty$.

The analytical difficulties of the Mumford-Shah functional arise from the need to balance smooth regions with discontinuities, the non-convexity due to the unknown edge set and the challenges in establishing lower-semicontinuity and compactness. Despite these complexities, various approximation methods have been developed, facilitating its application in practical image processing tasks. These methods enable the functional to be more tractable while retaining its ability to effectively segment and detect edges in images.

The lack of differentiability of the functional for a suitable norm does not allow us to use, as is classical, Euler–Lagrange equations. Moreover, the discretization of the unknown discontinuity set is a very complex task.

Existence and regularity of minimizing pairs (u, K) is not simple. In order to reduce the complexity of MS functional, special functions with bounded variation (SBV) were introduced by E. De Giorgi and L. Ambrosio [30], providing a cohesive framework for studying free discontinuity problems, which are prevalent in variational calculus and image processing. SBV functions are a subset of functions with bounded variation (BV), distinguished by the nature of their distributional derivatives. In the SBV framework, the Mumford-Shah functional is reformulated as

$$E_{MS} = \int_\Omega |u - u^\delta|^2 d\mathbf{x} + \alpha \int_\Omega |\nabla u|^2 d\mathbf{x} + \beta \mathcal{H}^{N-1}(S_u) \tag{7.5}$$

$u \in SBV(\Omega)$. Here S_u denotes the discontinuity set of u. The closed set K then can be recovered setting $K = \bar{S}_u$. But using the original functional (7.3) or its weak version (7.5) it is difficult to find smooth approximations and develop operational algorithms, since both functionals are non-convex.

Among the many proposals to solve this problem, the Ambrosio-Tortorelli approximation introduces a smooth auxiliary function v to approximate the edge set K. The key idea is to reformulate the Mumford-Shah functional as a sequence of functionals E_ε that can be minimized more easily, such that $E_\varepsilon \to \mathscr{J}_{\alpha,\beta,u^\delta}^{MS}(u, K)$ as $\varepsilon \to 0$ (on some kind of convergence). The Ambrosio-Tortorelli energy functional is defined as:

7.1 The Mumford-Shah Functional

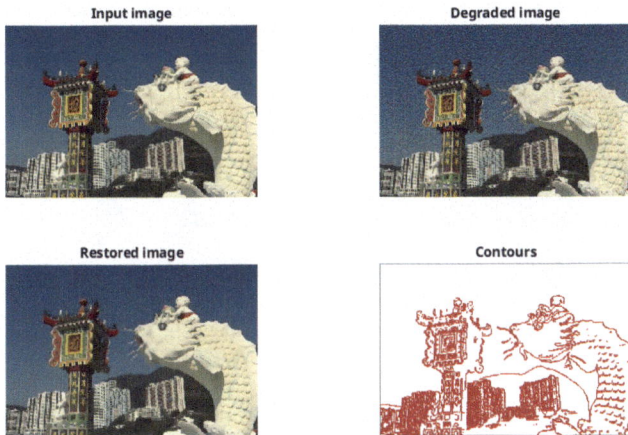

Fig. 7.1 Results for Mumford-Shah Denoising and segmentation

$$E_\varepsilon(u, v) = \int_\Omega (u - u^\delta)^2 d\mathbf{x} + \alpha \int_\Omega v^2 |\nabla u|^2 d\mathbf{x} + \beta \underbrace{\int_\Omega \left(\varepsilon |\nabla v|^2 + \frac{(1 - v^2)}{4\varepsilon} \right) d\mathbf{x}}_{Edge\,Penalty\,Term} \quad (7.6)$$

$\varepsilon |\nabla v|^2$ penalizes large gradients in v, promoting smooth transitions.

$\frac{(1-v^2)}{4\varepsilon}$ ensures that v stays close to 1 in smooth regions and close to 0 at edges, effectively approximating the edge set. Observe that for a more efficient approximation, the set K does not appear in E_ε.

The optimization of the Mumford-Shah functional in computer vision and image processing [2, 3, 16, 82] has evolved significantly, with various methods developed to address its inherent challenges. From variational methods and discrete approximations to convex relaxation [68] and proximal algorithms [25], each approach offers unique advantages and has contributed to the advancement of image analysis techniques. Practical implementations often involve a combination of several techniques to balance computational efficiency and solution quality [18]. Figure 7.1 shows an example obtained with the implementation of [41].

It is beyond the scope of this book to examine in detail these techniques but we can say, in general terms, that almost all optimization theories have been put at the service of image processing problems. The reader may find more information in [2, 25, 41, 82] and the extensive literature.

Example Although the initial idea of the Mumford-Shah model was image segmentation, it can also be used to eliminate noise. One way to achieve this is by assigning appropriate values to the regularization parameters. An illustration of this case can be found in [57], which use the piecewise constant Mumford-Shah model, also known as the Potts model in the discrete setting, originating in solid state physics [93]. The Potts model assumes that the underlying signal is constant

Fig. 7.2 Results for Mumford-Shah Denoising of speckle noise

between its discontinuities. The corresponding minimization problem is given by the functional

$$P_\gamma = \gamma \|\nabla u\|_0 + \|Au - u^\delta\|_2^2$$

where $\|\nabla u\|_0 = |\{i : u_i \neq u_{i+1}\}|$ denotes the number of jumps of the target variable u [123]. Figure 7.2 shows one result for image restoration with speckle noise. □

With over 7500 citations as of 2024, the Mumford-Shah functional [83] remains one of the most influential publications in image analysis. This seminal work, along with related publications, has spurred extensive research on discontinuity-preserving smoothing, piecewise-smooth approximations, and minimal partition problems.

7.2 Variational Level Set Methods

In many applications the boundaries between different media play a crucial role. These curves, often referred to as **interfaces**, represent the boundaries or surfaces that separate distinct regions or phases within a system. Examples include the interface between air and water in fluid dynamics, the boundaries between different materials in materials science, or the contours of organs in medical imaging.

Understanding how these interfaces evolve over time is essential for predicting the behavior of complex systems. These interfaces may deform, propagate, merge, split, or disappear over time, driven by the underlying dynamics or physical laws governing the system. These dynamic changes in the interfaces are often governed by the geometry of the interface itself or by the underlying physics of the problem, such as fluid flow, heat transfer, or surface tension (Fig. 7.3).

The level set method, introduced by Osher and Sethian [86], provides a powerful framework for addressing these challenges. By representing interfaces implicitly as the zero level set of a higher-dimensional function $\phi(\mathbf{x}, t)$, the level set method naturally handles topological changes in the interfaces. The evolution of the level

7.2 Variational Level Set Methods

Fig. 7.3 An example of Level sets evolution for a function $z = u(x, y)$

set function (LSF) $\phi(\mathbf{x}, t)$ is governed by partial differential equations (PDEs) that describe how the zero level set moves over time, capturing the dynamic behavior of the interfaces accurately and robustly. $\phi(\mathbf{x}, t)$ represents the interface $\boldsymbol{\gamma}(t)$ as the set where $\phi(\mathbf{x}, t) = 0$. The level set function at pseudo time t can be defined as follows,

$$\begin{cases} \phi(\mathbf{x}, t) > 0 & \mathbf{x} \in \Omega \\ \phi(\mathbf{x}, t) < 0 & \mathbf{x} \notin \bar{\Omega} \\ \phi(\mathbf{x}, t) = 0 & \mathbf{x} \in \partial\Omega = \boldsymbol{\gamma}(t) \end{cases} \quad (7.7)$$

In a variational scheme like (7.9), the main concept involves guiding the boundary $\boldsymbol{\gamma}(t)$ from its starting point towards the direction $-\nabla E$ of decreasing energy gradient. If the contour evolves along the normal $\boldsymbol{n}$ with a speed F then

$$\boldsymbol{\gamma}'(t) = F\boldsymbol{n}$$

Given that $\phi(x(s, t), y(s, t), t) = 0$ we can say by the chain rule

$$\frac{d}{dt}\phi(\boldsymbol{\gamma}(t), t) = \frac{\partial \phi}{\partial x}\frac{\partial x}{\partial t} + \frac{\partial \phi}{\partial y}\frac{\partial y}{\partial t} + \frac{\partial \phi}{\partial t}$$

$$= \nabla\phi\frac{\partial \boldsymbol{\gamma}}{\partial t} + \frac{\partial}{\partial t}$$

$$= \nabla\phi F \cdot \boldsymbol{n} + \frac{\partial \phi}{\partial t} = 0$$

and replacing with the formula $\boldsymbol{n} = \frac{\nabla\phi}{\|\nabla\phi\|}$, we obtain the evolution equation for ϕ

$$\frac{\partial \phi}{\partial t} = \|\nabla\phi\| F \quad (7.8)$$

It is possible to define some geometrical properties of the curve $\boldsymbol{\gamma}$ in terms of the level set function ϕ

Curvature at $(x, y) \in \gamma$

$$\kappa(x, y) = \frac{\partial}{\partial x}\left(\frac{\phi_x}{\|\nabla\phi\|}\right) + \frac{\partial}{\partial y}\left(\frac{\phi_x}{\|\nabla\phi\|}\right)$$

The Heaviside function $H(\phi) = \begin{cases} 1; \phi \geq 0 \\ 0; \phi < 0 \end{cases}$ and delta function $\delta(\phi) = H'(\phi)$ in the sense of distributions. The area enclosed by the curve is

$$A\{\phi \geq 0\} = \int_\Omega H(\phi(x, y)) dy dx$$

and the length of the curve

$$L\{\phi = 0\} = \int_\Omega \|\nabla H(\phi)\|$$

Variational methods for image segmentation using level set techniques are a powerful approach that formulate the segmentation problem as an energy minimization task. These methods can be classified based on the type of energy functional used, the nature of the image features they exploit, and their application to different segmentation scenarios. Two primary categories of level set techniques for image segmentation are Region-based and edge-based methods each with distinct approaches and characteristics.

Edge-based methods (geodesic active contour) use image gradient information to evolve the contour towards object boundaries. These methods focus on detecting and utilizing edges (discontinuities in image intensity) as the primary cues for segmentation.

Region-based methods (Chan-Vese model) utilize statistical properties of image regions (such as intensity homogeneity) to guide the contour evolution. These methods focus on partitioning the image into regions with similar characteristics.

7.3 Active Contours

Active contours are curve evolution techniques used for image segmentation. They are particularly useful for capturing object boundaries in images. Active contour models represent evolving curves that move towards object boundaries by minimizing an energy functional.

The conventional approach begins by initializing a curve, denoted as the **active contour** or **snake** near a contour within the image. The objective is to find permissible deformations of this contour that enable it to converge towards the

7.3 Active Contours

desired contour. The energy functional is designed to encapsulate image features, with the set of local minima representing the sought-after features within the image.

The concept of snakes was introduced in the seminal paper [61]. That is basically the approach we present here with the implementation given in [84] based on image gradient information in the following steps

- Start with an initial close curve $s \in [0, 1] \rightarrow \boldsymbol{\gamma}(s) = (x(s), y(s))$; so that $\boldsymbol{\gamma}(0) = \boldsymbol{\gamma}(1)$
- The curve deforms by minimizing an Energy functional $\mathscr{J}(\boldsymbol{\gamma}(s))$
- The curve stop in the boundaries of objects

The energy functional $\mathscr{T}_{snake}(\boldsymbol{\gamma})$ is composed of three terms from which we consider here only the first two terms such that:

$$\mathscr{T}_{snake} = \mathscr{T}_{int}(\boldsymbol{\gamma}) + \mathscr{T}_{ext}(\boldsymbol{\gamma})$$

where the internal energy is given by

$$\mathscr{T}_{int}(\boldsymbol{\gamma}) = \alpha \int_0^1 |\boldsymbol{\gamma}'|^2 ds + \beta \int_0^1 |\boldsymbol{\gamma}''|^2 ds \quad (7.9)$$

$|\boldsymbol{\gamma}'|$ is the membrane term and $\boldsymbol{\gamma}''$, the thin-plate term. There are several ways to choose the external potential energy

$$\mathscr{T}_{ext}(\boldsymbol{\gamma}) = w_1 \mathscr{T}_{line} + w_2 \mathscr{T}_{edge} + w_3 E_{term}$$

This expression consists of three functionals that guide the snake towards prominent features of the image. For example, if $E_{line} = \int_0^1 u(\boldsymbol{\gamma}(s)) ds$, the snake will be attracted to either light or dark lines depending on the sign of w_1. The snake is attracted to contours with large image gradients with $E_{edge} = -\int_0^1 |\nabla u(\boldsymbol{\gamma}(s))|^2 ds$. The third term could be based on the curvature as $E_{term} = \int_0^1 curv(G_\sigma * u) \boldsymbol{\gamma}(s) ds$ to detect terminations with high curvature. So a simplified version of the snake could be

$$\mathscr{T}_{int}(\boldsymbol{\gamma}) = \alpha \int_0^1 |\boldsymbol{\gamma}'|^2 ds + \beta \int_0^1 |\boldsymbol{\gamma}''|^2 ds - \lambda \int_0^1 |\nabla u(\boldsymbol{\gamma}(s))|^2 ds \quad (7.10)$$

The image energy attracts the snake to low-level features, such as brightness or edge data, aiming to select those with least contribution. The original formulation suggested that lines, edges and terminations could contribute to the energy function.

Example Note that this is a functional in the form $\mathscr{T}[\boldsymbol{\gamma}] = \mathscr{T}[(x(s), y(s))]$ in accordance with (4.8). Then for minimizing a simplified version of (7.10)

$$\mathscr{T}(\boldsymbol{\gamma}) = \alpha \int_0^1 |\boldsymbol{\gamma}'|^2 ds + \beta \int_0^1 |\boldsymbol{\gamma}''|^2 ds \int_0^1 E(\boldsymbol{\gamma}(s)) ds$$

we apply (4.9). Given that

$$\mathcal{T} = \int (\alpha x'^2 + \beta x''^2)ds + \int (\alpha y'^2 + \beta y''^2)ds + \int E(x(s), y(s))ds$$

For the first functional we have the Euler equation

$$-\frac{d}{ds}\left(\alpha(s)\frac{dx}{ds}\right) + \frac{d^2}{ds^2}\left(\beta(s)\frac{dx^2}{ds^2}\right) = 0$$

and a similar equation in $y(s)$ for the second functional. As for the functional $\int E(x(s), y(s))ds$, applying (4.9) their Euler equations are

$$\frac{\partial E}{\partial x} - \underbrace{\frac{d}{ds}\frac{\partial E}{\partial x'}}_{=0} = 0; \qquad \frac{\partial E}{\partial y} - \underbrace{\frac{d}{ds}\frac{\partial E}{\partial y'}}_{=0} = 0$$

thus summing up all the equations

$$-\frac{d}{ds}\left(\alpha(s)\frac{dx}{ds}\right) + \frac{d^2}{ds^2}\left(\beta(s)\frac{dx^2}{ds^2}\right) + \frac{\partial E}{\partial x} = 0$$

$$-\frac{d}{ds}\left(\alpha(s)\frac{dy}{ds}\right) + \frac{d^2}{ds^2}\left(\beta(s)\frac{dy^2}{ds^2}\right) + \frac{\partial E}{\partial y} = 0$$

which can be solved numerically. □

Using a positive and decreasing function g (with $\lim_{z\to\infty}$) can be defined a more general edge detector. A well-known choice is $g(t) = \frac{1}{1+t^p}$, $p \geq 1$, to obtain

$$g(\|\nabla u(\mathbf{x})\|) = \frac{1}{1 + \|\nabla(G_\sigma * u)(\mathbf{x})\|^p}; p \geq 1$$

where G_σ is the Gaussian kernel (3.47), in homogeneous regions the function $g(\|\nabla u(\mathbf{x})\|)$ is strictly positive, and close to zero on edges. Then the snake model without elastic term becomes

$$\mathcal{T}(\boldsymbol{\gamma}) = \int_0^1 |\boldsymbol{\gamma}'(s)|^2 ds + \lambda \int_0^1 g(|\nabla u(\boldsymbol{\gamma}(s))|)^2 ds$$

When considering the representation of the curve, active contours are typically divided into two categories. **Parametric snakes** use an explicit parametric representation of the contour, evolving control points directly to minimize an energy functional. They are simple to implement but struggle with topological changes. **Geometric snakes** use an implicit representation via level set functions, evolving the zero level set to minimize a geometric energy functional. They handle topological changes well but are more complex and computationally demanding.

7.3 Active Contours

Geodesic Active Contour (GAC) Models, an extension of snakes, were introduced by Caselles et al. in [14] to improve the segmentation of objects with irregular shapes. These models incorporate geodesic distances into the energy functional to guide the contour more effectively.

$$\inf_{\gamma} \mathscr{T}(\gamma) = \int_0^1 \|\gamma'(s)\| \cdot g(\|\nabla u(\gamma(s))\|) ds \tag{7.11}$$

These evolving contours can naturally split and merge, enabling the simultaneous detection of multiple objects and their boundaries, both inside and outside. The approach links active contours to the computation of geodesics, or minimal distance curves, within a Riemannian space defined by the image's content.

The evolution of the curve γ is governed by a partial differential equation (PDE), derived from the Euler-Lagrange equations associated with the energy functional (7.11)

$$\frac{\partial \gamma}{\partial t} = g\kappa \boldsymbol{n} - (\nabla g \cdot \boldsymbol{n})\boldsymbol{n}$$

where $\boldsymbol{n}$ is the inward unit normal vector to the curve, $g := g(|\nabla u|)$ and κ is the curvature of the curve.

In the level set framework (7.11) can be expressed as

$$\inf_{\phi} \mathscr{T}[\phi] = \int_{\Omega} g(|\nabla u(\mathbf{x})|) |DH(\phi)| + \nu \int_{\Omega} g(|\nabla u|) H(\phi) d\mathbf{x}$$

where the second term is a so called *balloon force term* which refers to an additional term in the energy functional or the evolution equation that drives the contour outward or inward, effectively "inflating" or "deflating" it like a balloon. This term is introduced to counteract issues such as weak or broken edges in the image that might otherwise prevent the contour from fully enclosing an object. Replacing with smooth approximations H_ε and δ_ε

$$\mathscr{T}[\phi] = \int_{\Omega} g(|\nabla u(\mathbf{x})|) \delta_\varepsilon(\phi(\mathbf{x})) |\nabla \phi(\mathbf{x})| d\mathbf{x} + \nu \int_{\Omega} g(|\nabla u|) H_\varepsilon(\phi) d\mathbf{x} \tag{7.12}$$

Calculating the first variation of (7.12) [119], we have

$$\frac{\partial}{\partial \varepsilon} \mathscr{T}[\Phi + \varepsilon \Psi] \bigg|_{\varepsilon=0} = -\int_{\Omega} \delta_\varepsilon(\Phi) div\left(g \frac{\nabla \Phi}{|\nabla \Phi|}\right) \Psi d\mathbf{x} + \nu \int_{\Omega} g \delta_\varepsilon(\Phi) \Psi d\mathbf{x}$$
$$+ \int_{\partial \Omega} g \frac{\delta_\varepsilon(\Phi)}{|\nabla \Phi|} \frac{\partial \Phi}{\partial \boldsymbol{n}} \Psi d\sigma$$

$$\tag{7.13}$$

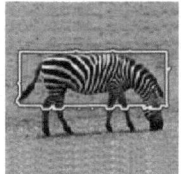 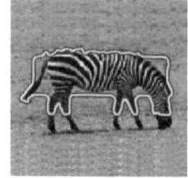 

Fig. 7.4 GAC Geodesic Active Contours Results. 5, 100 and 400 iterations from left to right

so the Euler equation is

$$-\delta_\varepsilon(\Phi) div\left(g\frac{\nabla\Phi}{|\nabla\Phi|}\right) + \nu g\delta_\varepsilon(\Phi) = 0$$

with boundary conditions $g\dfrac{\delta_\varepsilon(\Phi)}{|\nabla\Phi|}\dfrac{\partial\Phi}{\partial \boldsymbol{n}} = 0$ on $\partial\Omega$

Example Since their introduction, several variants of GAC have been developed to address various challenges and improve performance. Each of these variants of geodesic active contours introduces modifications to the original model to address specific challenges or improve segmentation performance. By combining edge information, region information, balloon forces, and shape priors, these methods provide a versatile toolkit for robust image segmentation.

As an instance of the many variations of GAC model, we present here (Fig. 7.4) some results obtained running the code in [103].

7.4 The Chan-Vese Model for Image Segmentation

Chan-Vese Model [119] The Chan-Vese model simplifies the Mumford-Shah model for binary segmentation (two regions: object and background). It is designed for piecewise constant segmentation and is particularly effective in scenarios where objects have relatively homogeneous intensity values [20, 21, 23].

Compared to the piecewise constant Mumford–Shah model, the key differences with the Chan–Vese model are an additional term penalizing the enclosed area and a further simplification that u is allowed to have only two values,

$$u(x) = \begin{cases} c_1; & \boldsymbol{x} \text{ inside } \boldsymbol{\gamma} \\ c_2; & \boldsymbol{x} \text{ outside } \boldsymbol{\gamma} \end{cases}$$

where $\boldsymbol{\gamma}$ is the boundary of a closed set and c_1, c_2 are the values of u respectively inside and outside of $\boldsymbol{\gamma}$. Therefore, the unknown variable $\boldsymbol{\gamma}$ is replaced by ϕ and the new energy, $\mathscr{T}(\phi, c_1, c_2)$ becomes:

7.4 The Chan-Vese Model for Image Segmentation

$$\mathcal{F}(\phi, c_1, c_2) = \mu \cdot \{\phi = 0\} + \nu \cdot area\{\phi \geq 0\}$$
$$+ \lambda_1 \int_{\phi \geq 0} |u_0 - c_1|^2 dydx + \lambda_2 \int_{\phi < 0} |u_0 - c_2| dydx \quad (7.14)$$

The terms in $\mathcal{F}$ are expressed as

$$\int_{\phi \geq 0} |u_0 - c_1|^2 dydx = \int_{\Omega} |u_0 - c_1|^2 H(\phi) dydx$$

$$\int_{\phi < 0} |u_0 - c_2|^2 dydx = \int_{\Omega} |u_0 - c_2|^2 (1 - H(\phi)) dydx$$

and the functional can be written as

$$\mathcal{F}(\phi, c_1, c_2) = \mu \int_{\Omega} \delta(\phi) \|\nabla \phi\| \nu \int_{\Omega} H(\phi) dydx$$
$$+ \lambda_1 \int_{\Omega} |u_0 - c_1|^2 H(\phi) dydx + \lambda_2 \int_{\Omega} |u_0 - c_2|^2 (1 - H(\phi)) dydx$$

Considering ϕ fixed and minimizing the energy $\mathcal{F}(\phi, c_1, c_2)$ with respect to the constants c_1, c_2, the constants can be expressed in terms of ϕ

$$c_1(\phi) = \frac{\int_{\Omega} u_0 H(\phi) dydx}{\int_{\Omega} H(\phi(x, y)) dydx} \quad (7.15)$$

$$c_2(\phi) = \frac{\int_{\Omega} u_0 (1 - H(\phi(x, y))) dydx}{\int_{\Omega} (1 - H(\phi(x, y))) dydx} \quad (7.16)$$

Taking c_1, c_2 as constants, the gradient of the functional is

$$\nabla \mathcal{F} = \delta(\phi) \left[\mu div \left(\frac{\nabla \phi}{\|\nabla \phi\|} \right) - \nu - \lambda_1 (u_0 - c_1)^2 + \lambda_2 (u_0 - c_2)^2 \right]$$

The numerical implementation requires regularized versions of $\delta(\phi)$ and $H(\phi)$; for which we can find several implementations. For example [131] uses

$$H_\alpha = \begin{cases} 1, & x > \alpha \\ 0, & x < -\alpha \\ \frac{1}{2}\left[1 + \frac{x}{\alpha} + \frac{1}{\pi}\sin(\frac{\pi x}{\alpha})\right], & |x| \leq \alpha \end{cases}$$

$$\delta_\alpha(x) = \frac{d}{dx} H_\alpha(x) = \begin{cases} 0, & |x| > \alpha \\ \frac{1}{2\alpha}\left[1 + \cos(\frac{\pi x}{\alpha})\right], & |x| \leq \alpha \end{cases}$$

Fig. 7.5 Plots for δ_ε and H_ε with $\varepsilon = 1$

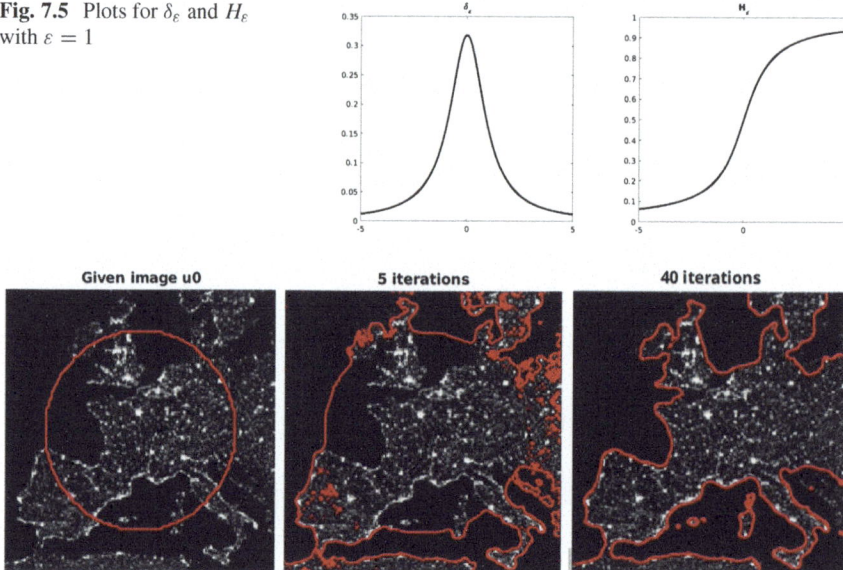

Fig. 7.6 Chan-Vese model for image segmentation

and [22] uses instead the following C^∞ regularized versions (Fig. 7.5)

$$H_\varepsilon(x) = \frac{1}{2}(1 + \frac{2}{\pi}\arctan(\frac{x}{\varepsilon})); \qquad \delta_\varepsilon = \frac{1}{\pi} \cdot \frac{\varepsilon}{\varepsilon^2 + x^2}$$

Figure 7.6, shows one result of the Chan-Vese model for image segmentation based on [22] and [69]

References

1. Al-Gwaiz MA (1992) Theory of distributions. CRC Press
2. Aubert G, Kornprobst P (2010) Mathematical problems in image processing. Partial differential equations and the calculus of variations, 2nd edn. Springer
3. Bar L, Chan T, Chung G, Jung M, Kiryati N, Sochen N, Vese LA (2015) Mumford and Shah model and its applications to image segmentation and image restoration. In: Handbook of mathematical methods in imaging: Volume 1, 2nd edn., pp. 1539–1597. Springer New York
4. Basdevant JL (2023) Variational principles in physics. Springer Nature
5. Beck A (2014) Introduction to nonlinear optimization: Theory, algorithms, and applications with MATLAB. Society for Industrial and Applied Mathematics
6. Beck A (2017) First-order methods in optimization. Society for Industrial and Applied Mathematics
7. Bertero M, Boccacci P, De Mol C (2021) Introduction to inverse problems in imaging. CRC Press
8. Blake A, Zisserman A (1987) Visual reconstruction. MIT Press
9. Blanchard P, Brüning E (2012) Variational methods in mathematical physics: a unified approach. Springer Science and Business Media
10. Blum EK (1972) Numerical analysis and computation:Theory and Practice. Addison-Wesley Pub. Co.
11. Bredies K, Lorenz D (2018) Mathematical image processing. Springer International Publishing, Cham, pp. 1–469
12. Bredies K, Kunisch K, Pock T (2010) Total generalized variation. SIAM J. Imaging Sci. 3(3):492–526
13. Burden RL, Faires JD, Burden AM (2015) Numerical analysis. Cengage learning
14. Caselles V, Kimmel R, Sapiro G (1997) Geodesic active contours. Int J Comput Vis 22:61–79
15. Cassel KW (2013) Variational methods with applications in science and engineering. Cambridge University Press
16. Chambolle A (1999) Finite-differences discretizations of the Mumford-Shah functional. ESAIM: Math Model Numer Anal 33(2):261–288
17. Chambolle A (2004) An algorithm for total variation minimization and applications. J Math Imaging Vis 20:89–97
18. Chambolle A, Pock T (2016) An introduction to continuous optimization for imaging. Acta Numer 25:161–319
19. Chambolle A, Caselles V, Cremers D, Novaga M, Pock T (2010) An introduction to total variation for image analysis. In: Theoretical foundations and numerical methods for sparse recovery, vol 9(263–340), p 227

20. Chan TF, Shen J (2005) Variational image inpainting. Commun Pure Appl Math 58(5):579–619
21. Chan TF, Shen J (2005) Image processing and analysis: variational, PDE, wavelet, and stochastic Methods. SIAM
22. Chan TF, Vese LA (2001) Active contours without edges. IEEE Trans Image Process 10(2):266–277
23. Chan TF, Shen J, Vese L (2003) Variational PDE models in image processing. Not AMS 50(1):14–26
24. Chen K (2013) Introduction to variational image-processing models and applications. Int J Comput Math 90(1), 1–8
25. Chen K, Schönlieb CB, Tai XC, Younes L (eds) (2023) Handbook of mathematical models and algorithms in computer vision and imaging: mathematical imaging and vision. Springer Nature
26. Cheney EW, Kincaid DR (2012) Numerical mathematics and computing. Cengage Learning
27. Courant R, Hilbert D (1953) Methods of mathematical physics. Interscience Publishers
28. Dacorogna B (2007) Direct methods in the calculus of variations, vol 78. Springer Science and Business Media
29. Dacorogna B (2014) Introduction to the calculus of variations. World Scientific Publishing Company
30. De Giorgi E, Ambrosio L (1989) New functionals in calculus of variations. In: Nonsmooth optimization and related topics. Springer US, Boston, MA, pp 49–59
31. Dirac PAM (1927) The physical interpretation of the quantum dynamics. Proc R Soc Lond Ser A 113(765):621–641
32. Dirac PAM (1958) The principles of quantum mechanics, 4th edn. Oxford University Press
33. Do Carmo MP (2016) Differential geometry of curves and surfaces: revised and updated second edition. Courier Dover Publications
34. Duchon J (1977) Splines minimizing rotation-invariant semi-norms in Sobolev spaces. In: Constructive theory of functions of several variables: Proceedings of a conference held at Oberwolfach April 25–May 1, 1976. Springer Berlin Heidelberg, pp 85–100
35. Duistermaat JJ, Kolk JAC (2010) Distributions: theory and applications. Birkhäuser Boston
36. Elsgolc LD (2012) Calculus of variations. Courier Corporation
37. Engel E, Dreizler RM (2011) Density functional theory: an advanced course series. Theoretical and mathematical physics. Springer
38. Engl HW, Hanke M, Neubauer A (1996) Regularization of inverse problems, vol 375. Springer Science and Business Media
39. Fasshauer GE (2007) Meshfree approximation methods with MATLAB, vol 6. World Scientific
40. Fleming W (2012) Functions of several variables. Springer Science Business Media
41. Foare M, Pustelnik N, Condat L (2019) Semi-linearized proximal alternating minimization for a discrete Mumford–Shah model. IEEE Trans Image Process 29:2176–2189
42. Folland GB (1999) Real analysis: modern techniques and their applications, vol 40. John Wiley and Sons
43. Folland GB (2009) Fourier analysis and its applications, vol 4. American Mathematical Society
44. Fox C (1987) An introduction to the calculus of variations. Courier Corporation
45. Gel'fand IM, Shilov GE (2016) Generalized functions, volume 2: Spaces of fundamental and generalized functions, vol 261. American Mathematical Society
46. Gelfand IM, Silverman RA (2000) Calculus of variations. Courier Corporation
47. Geman S, Geman D (1984) Stochastic relaxation, Gibbs distributions, and the Bayesian restoration of images. IEEE Trans Pattern Anal Mach Intell (6):721–741
48. Giaquinta M, Hildebrandt S (2013) Calculus of variations II, vol 311. Springer Science and Business Media
49. Gilboa G (2018) Nonlinear eigenproblems in image processing and computer vision. Springer International Publishing

50. Giusti E, Williams GH (1984) Minimal surfaces and functions of bounded variation, vol 80. Birkhäuser, Boston, pp xii+-240
51. Goldstine HH (2012) A history of the calculus of variations from the 17th through the 19th century, vol 5. Springer Science Business Media
52. Golomb M (1935) Zur Theorie der nichtlinearen Integralgleichungen, Integralgleichungssysteme und allgemeinen Funktionalgleichungen. Math Z 39(1):45–75
53. Greiner G, Loos J, Wesselink W (1996, August) Data dependent thin plate energy and its use in interactive surface modeling. In: Computer graphics forum, vol 15, no 3. Blackwell Science, Edinburgh, pp 175–185
54. Grimson WEL (1983) An implementation of a computational theory of visual surface interpolation. Comput Vis Graph Image Process 22(1):39–69
55. Grimson WEL, Grimson WEL (1981) From images to surfaces: A computational study of the human early visual system, vol 4. MIT Press, Cambridge, MA
56. Han W, Atkinson KE (2009) Theoretical numerical analysis: A functional analysis framework. Springer New York
57. Hohm K, Storath M, Weinmann A (2015) An algorithmic framework for Mumford–Shah regularization of inverse problems in imaging. Inverse Probl. 31(11):115011
58. Jähne B (2005) Digital image processing. Springer Science and Business Media
59. Kalman RE (1963) The theory of optimal control and the calculus of variations. Math Optim Tech 309:329
60. Kanwal RP (2004) Generalized functions: theory and applications. Springer Science and Business Media
61. Kass M, Witkin A, Terzopoulos D (1988) Snakes: Active contour models. Int J Comput Vis 1(4):321–331
62. Kesavan S (2023) Functional analysis, vol 52. Springer Nature
63. Kincaid DR, Cheney EW (2009) Numerical analysis: mathematics of scientific computing, vol 2. American Mathematical Society
64. Komzsik L (2019) Applied calculus of variations for engineers. CRC Press
65. Kravvaritis DC, Yannacopoulos AN (2020) Variational methods in nonlinear analysis: with applications in optimization and partial differential equations. Walter De Gruyter
66. Kress R (1998) Numerical analysis, vol 181. Springer Science and Business Media
67. Lanczos C (1970) The variational principles of mechanics, 4th edn. University of Toronto Press (Dover 1986)
68. Lanza A, Morigi S, Selesnick IW, Sgallari F (2022) Convex non-convex variational models. In: Handbook of mathematical models and algorithms in computer vision and imaging: mathematical imaging and vision. Springer International Publishing, Cham, pp 1–57
69. Li C, Xu C, Gui C, Fox MD (2005, June) Level set evolution without re-initialization: a new variational formulation. In: 2005 IEEE computer society conference on computer vision and pattern recognition (CVPR'05), vol 1. IEEE, pp 430–436
70. Linz P (2019) Theoretical numerical analysis. Courier Dover Publications
71. Lipschutz MM (1969) Schaum's outline of differential geometry. McGraw Hill Professional
72. Love AEH (2013) A treatise on the mathematical theory of elasticity. Cambridge University Press
73. Lu W, Duan J, Qiu Z, Pan Z, Liu RW, Bai L (2016) Implementation of high-order variational models made easy for image processing. Math Methods Appl Sci 39(14):4208–4233
74. Luenberger DG (1969) Optimization by vector space methods. John Wiley and Sons
75. Lützen J (2012) The prehistory of the theory of distributions, vol 7. Springer Science and Business Media
76. Ma J, Wang D, Wang XP, Yang X (2021) A characteristic function-based algorithm for geodesic active contours. SIAM J Imaging Sci 14(3):1184–1205s
77. Marr D (2010) Vision: A computational investigation into the human representation and processing of visual information. MIT Press
78. Mathews JH, Fink KD (2004) Numerical methods using MATLAB, vol 4. Pearson Prentice Hall, Upper Saddle River, NJ

79. Meinguet J (1979) Multivariate interpolation at arbitrary points made simple. Z Angew Math Phys ZAMP 30(2):292–304
80. Mitiche A, Ayed IB (2010) Variational and level set methods in image segmentation, vol 5. Springer Science Business Media
81. Modersitzki J (2003) Numerical methods for image registration. OUP Oxford
82. Morel JM, Solimini S (1995) Variational methods in image segmentation. Birkhäuser, Boston
83. Mumford D, Shah J (1989) Optimal approximations by piecewise smooth functions and associated variational problems. Commun Pure Appl Math 42:577–685
84. Nixon M, Aguado A (2019) Feature extraction and image processing for computer vision. Academic Press
85. Oprea J (2024) Differential geometry and its applications, vol 59. American Mathematical Society
86. Osher S, Sethian JA (1988) Fronts propagating with curvature-dependent speed: Algorithms based on Hamilton-Jacobi formulations. J Comput Phys 79(1):12–49
87. Paragios N, Chen Y, Faugeras OD (eds) (2006) Handbook of mathematical models in computer vision. Springer Science and Business Media
88. Parker JR (2010) Algorithms for image processing and computer vision. John Wiley and Sons
89. Pathak HK (2018) An introduction to nonlinear analysis and fixed point theory. Springer
90. Peypouquet J (2015) Convex optimization in normed spaces: theory, methods and examples. Springer
91. Peyre G. Matlab's Tours. http://www.numerical-tours.com/matlab/. Accessed 17 Aug 2024
92. Poggio T, Torre V, Koch C (1987) Computational vision and regularization theory. In: Readings in computer vision, pp 638–643
93. Potts RB (1952, January) Some generalized order-disorder transformations. In: Mathematical proceedings of the Cambridge philosophical society, vol 48, no 1. Cambridge University Press, pp 106–109
94. Powell MJD (1981) Approximation theory and methods. Cambridge University Press
95. Reddy JN (1986) Applied functional analysis and variational methods. McGraw-Hill Book Company
96. Reddy JN (2002) Energy principles and variational methods in applied mechanics, 2nd edn. John Wiley and Sons
97. Riesenhuber M, Poggio T (2000) Models of object recognition. Nat Neurosci 3(11):1199–1204
98. Rindler F (2018) Calculus of variations, vol 5. Springer, Berlin
99. Rojo A, Bloch A (2018) The principle of least action: History and physics. Cambridge University Press
100. Rudin W (1976) Principles of mathematical analysis, vol 3. McGraw-Hill, New York
101. Rudin W (1991) Functional analysis. McGraw-Hill, New York
102. Rudin LI, Osher S, Fatemi E (1992) Nonlinear total variation based noise removal algorithms. Physica D Nonlinear Phenomena 60(1–4):259–268
103. Sandhu R, Georgiou T, Tannenbaum A (2008, March) A new distribution metric for image segmentation. In: Medical imaging 2008: image processing, vol 6914. SPIE, pp 40–48
104. Scherzer O (ed) (2010) Handbook of mathematical methods in imaging. Springer Science and Business Media
105. Scherzer O, Grasmair M, Grossauer H, Haltmeier M, Lenzen F (2009) Variational methods in imaging, vol 167. Springer Science+ Business Media LLC
106. Schoenberg IJ (1964, June) On interpolation by spline functions and its minimal properties. In: On approximation theory/Über Approximationstheorie: Proceedings of the conference held in the Mathematical Research Institute at Oberwolfach, Black Forest, August 4–10, 1963/Abhandlungen zur Tagung im Mathematischen Forschungsinstitut Oberwolfach, Schwarzwald, vom 4.–10. August 1963. Springer Basel, Basel, pp 109–129
107. Schwartz L (1951) Théorie des distributions. Hermann
108. Schwartz L (2008) Mathematics for the physical sciences. Dover Publications
109. Simmons GF (2016) Differential equations with applications and historical notes. CRC Press

110. Smith DR (1998) Variational methods in optimization. Courier Corporation
111. Stakgold I (2000) Boundary value problems of mathematical physics: Volume 1,2. Society for Industrial and Applied Mathematics
112. Strichartz RS (2003) A guide to distribution theory and Fourier transforms. World Scientific Publishing Company
113. Tapia RA, Trosset MW (1994) An extension of the Karush-Kuhn-Tucker necessity conditions to infinite programming. SIAM review 36(1):1–17
114. Tapia RA (1971) The differentiation and integration of nonlinear operators. In: Nonlinear functional analysis and applications. Academic Press, pp 45–101
115. Tikhonov AN, Arsenin VY (1977) Solutions of ill-posed problems. VH Winston and Sons
116. Timoshenko SP, Goodier JN (1934) Theory of elasticity. McGraw-Hill Book Company
117. Troutman JL (2012) Variational calculus and optimal control: optimization with elementary convexity. Springer Science Business Media
118. Vaĭnberg MM, Kantorovich LV, Akilov GP, Feinstein A (1964) Variational methods for the study of nonlinear operators. Holden-Day
119. Vese LA, Le Guyader C (2015) Variational methods in image processing. CRC Press
120. Vladimirov VS (1976) Generalized functions in mathematical physics. Moscow Izdatel Nauka
121. Vogel CR (2002) Computational methods for inverse problems. Society for Industrial and Applied Mathematics
122. Wahba G (1990) Spline models for observational data. Society for Industrial and Applied Mathematics
123. Weinmann A, Storath M (2015) Iterative Potts and Blake–Zisserman minimization for the recovery of functions with discontinuities from indirect measurements. Proc R Soc A Math Phys Eng Sci 471(2176):20140638
124. Weinstock R (1974) Calculus of variations: with applications to physics and engineering. Courier Corporation
125. Willem M (2023) Functional analysis: Fundamentals and applications. Springer Nature
126. Woods R, Gonzalez RC (2008) Digital image processing. Pearson Education India
127. Zeidler E (2006) Quantum field theory I: Basics in mathematics and physics: A bridge between mathematicians and physicists. Springer, Berlin
128. Zeidler E (2013) Nonlinear functional analysis and its applications: III: Variational methods and optimization. Springer Science and Business Media
129. Zeidler E (2013) Nonlinear functional analysis and its applications: IV: Applications to mathematical physics. Springer Science and Business Media
130. Zemanian AH (1987) Distribution theory and transform analysis: an introduction to generalized functions, with applications. Courier Corporation
131. Zhao HK, Chan T, Merriman B, Osher S (1996) A variational level set approach to multiphase motion. J Comput Phys 127(1):179–195

Index

A
Active contours, 134
Additive noise, 92
Adjoint operator, 40

B
Banach space, 37
Biharmonic operator, 87, 120
Bounded linear operator, 42
Bounded variation, 32

C
Cauchy sequence, 35
Chan-Vese model, 134
Chan-Vese segmentation, 139
Complete space, 37
Continuous function, 33
Continuous linear functional, 44
Convergence, 33
Convex functional, 102
Convexity, 102
Convex set, 31
Convolution, 59
Cubic spline, 112
Curvature, 95, 110, 134

D
Deblurring, 8, 90, 92
Denoising, 8, 11
Derivative, 19
Dirac delta, 46, 90, 116, 134, 139
Directional derivative, 19
Dirichlet kernel, 58

Discrete convolution, 69
Distribution, 52
Distributional derivative, 52, 55
Distributions, 46
Divergence, 23
Divided differences, 108
Duchon, 12

E
Edge-based methods, 134
Edge detection, 23
Energy functional, 12
Euclidian norm, 32
Euler-Lagrange equation, 2, 3, 5, 77
Extremum, 23

F
First fundamental form, 85
First variation, 77
Fourier, 56
Fourier transform, 62
Franke's function, 122
Free discontinuity problems, 127
Functional gradient, 2, 77
Fundamental lemma, 50
Fundamental solutions, 66

G
Gateaux derivative, 75
Gaussian curvature, 95
Gaussian kernel, 71

Geodesic active contour, 136
Geometric snake, 136
Gradient, 19, 77
Gradient algorithm, 2
Gradient descent, 25
Gradient descent algorithm, 88
Green's function, 117

H
Heaviside function, 46, 134, 139
Hessian, 23
Hilbert space, 38

I
Image segmentation, 129
Inner product, 15
Integration by parts, 28, 46, 50
Interpolation matrix, 114
Inverse Fourier Transform, 71
Inverse problem, 90, 92

K
Kernel, 58

L
Lagrange polynomial, 108
Laplacian operator, 23, 68
Lebesgue integral, 37
Level set evolution, 134
Level sets, 132
Linear combination, 30
Linear functional, 44
Linear space, 29

M
Mean curvature, 95
Membrane, 95
Minimal surfaces, 85, 87
MS denoising, 131
MS model, 127
MS segmentation, 131
Mumford-Shah (MS), 11, 127

N
Noise, 92
Norm, 32
Normed space, 32

O
Open ball, 16
Open set, 33
Orthogonal set, 38
Osher, 132

P
Parametric snake, 136
Plate, 95
Polynomial interpolation, 107
Principal curvatures, 95

Q
Quadratic form, 23

R
Region-based methods, 134
Regularization, 11, 90, 115
   methods, 92
   parameter, 115, 122
Riesz representation theorem, 45
Rudin-Osher-Fatemi (ROF) model, 11

S
Scattered data interpolation, 120
Second fundamental form, 86
Second variation, 82
Smoothing splines, 115
Snakes, 134
Sobolev deconvolution, 97
Sobolev regularization, 94
Sobolev space, 56
Strong convexity, 102
Subspace, 30
Surface interpolation, 107
SVD, 17

T
Taylor theorem, 23
Test function, 50
Thin plate, 95
Thin plate functional, 87
Thin plate spline, 12, 120
Tikhonov regularization, 94
Total variation (TV), 11, 94, 98
TV deconvolution, 102

V
Variation, 32
Variational calculus, 1, 10
Variational derivative, 2, 78
Variational regularization, 93
Variational spline, 110

W
Weak derivative, 51
Weak form, 77
Well-posed, 92

SPRINGER NATURE

GPSR Compliance

The European Union's (EU) General Product Safety Regulation (GPSR) is a set of rules that requires consumer products to be safe and our obligations to ensure this.

If you have any concerns about our products, you can contact us on ProductSafety@springernature.com

In case Publisher is established outside the EU, the EU authorized representative is:

Springer Nature Customer Service Center GmbH
Europaplatz 3
69115 Heidelberg, Germany

The manufacturer's authorised representative in the EU is Springer Nature Customer Service Centre GmbH, Europaplatz 3, 69115 Heidelberg, Germany. If you have any concerns regarding our products, please contact ProductSafety@springernature.com

Printed and bound by CPI Group (UK) Ltd, Croydon, CR0 4YY

26/03/2026

02078967-0006